LOW POWER UWB CMOS RADAR SENSORS

## ANALOG CIRCUITS AND SIGNAL PROCESSING SERIES

***Consulting Editor*: Mohammed Ismail. *Ohio State University***

*Titles in Series:*

**LOW POWER UWB CMOS RADAR SENSORS**
Paulino, Nuno, Goes, João, Steiger Garção, Adolfo
ISBN: 978-1-4020-8409-6

**THE GM/ID DESIGN METHODOLOGY FOR CMOS ANALOG LOW POWER INTEGRATED CIRCUITS**
Jespers, Paul G.A.
ISBN-10: 0-387-47100-6

**SUBSTRATE NOISE COUPLING IN RFICS**
Helmy, Ahmed, Ismail, Mohammed
ISBN: 978-1-4020-8165-1

**CIRCUIT AND INTERCONNECT DESIGN FOR HIGH BIT-RATE APPLICATIONS**
Veenstra, Hugo, Long, John R.
ISBN: 978-1-4020-6882-9

**HIGH-RESOLUTION IF-TO-BASEBAND SIGMADELTA ADC FOR CAR RADIOS**
Silva, Paulo G.R., Huijsing, Johan H.
ISBN: 978-1-4020-8163-7

**MULTI-BAND RF FRONT-ENDS WITH ADAPTIVE IMAGE REJECTION A DECT/BLUETOOTH CASE STUDY**
Vidojkovic, V., van der Tang, J., Leeuwenburgh, A., van Roermund, A.H.M.
ISBN: 978-1-4020-6533-0

**SILICON-BASED RF FRONT-ENDS FOR ULTRA WIDEBAND RADIOS**
Safarian, Aminghasem, Heydari, Payam
ISBN: 978-1-4020-6721-1

**DESIGN OF HIGH VOLTAGE XDSL LINE DRIVERS IN STANDARD CMOS**
Serneels, Bert, Steyaert, Michiel
ISBN: 978-1-4020-6789-1

**BASEBAND ANALOG CIRCUITS FOR SOFTWARE DEFINED RADIO**
Giannini, Vito, Craninckx, Jan, Baschirotto, Andrea
ISBN: 978-1-4020-6537-8

**HIGH-LEVEL MODELING AND SYNTHESIS OF ANALOG INTEGRATED SYSTEMS**
Martens, Ewout S.J., Gielen, Georges
ISBN: 978-1-4020-6801-0

**CMOS MULTI-CHANNEL SINGLE-CHIP RECEIVERS FOR MULTI-GIGABIT OPT...**
Muller, P., Leblebici, Y.
ISBN 978-1-4020-5911-7

**ANALOG-BASEBAND ARCHITECTURES AND CIRCUITS FOR MULTISTANDARD AND LOW-VOLTAGE WIRELESS TRANSCEIVERS**
Mak, Pui In, U, Seng-Pan, Martins, Rui Paulo
ISBN: 978-1-4020-6432-6

**FULL-CHIP NANOMETER ROUTING TECHNIQUES**
Ho, Tsung-Yi, Chang, Yao-Wen, Chen, Sao-Jie
ISBN: 978-1-4020-6194-3

**ANALOG CIRCUIT DESIGN TECHNIQUES AT 0.5V**
Chatterjee, S., Kinget, P., Tsividis, Y., Pun, K.P.
ISBN-10: 0-387-69953-8

**SWITCHED-CAPACITOR TECHNIQUES FOR HIGH-ACCURACY FILTER AND ADC...**
Quinn, P.J., Roermund, A.H.M.v.
ISBN 978-1-4020-6257-5

**LOW-FREQUENCY NOISE IN ADVANCED MOS DEVICES**
von Haartman, M., Östling, M.
ISBN 978-1-4020-5909-4

**ULTRA LOW POWER CAPACITIVE SENSOR INTERFACES**
Bracke, W., Puers, R. (et al.)
ISBN 978-1-4020-6231-5

**BROADBAND OPTO-ELECTRICAL RECEIVERS IN STANDARD CMOS**
Hermans, C., Steyaert, M.
ISBN 978-1-4020-6221-6

**CMOS SINGLE CHIP FAST FREQUENCY HOPPING SYNTHESIZERS FOR WIRELESS MULTI-GIGAHERTZ APPLICATIONS**
Bourdi, Taoufik, Kale, Izzet
ISBN: 978-1-4020-5927-8

**CMOS CURRENT-MODE CIRCUITS FOR DATA COMMUNICATIONS**
Yuan, Fei
ISBN: 0-387-29758-8

# LOW POWER UWB CMOS RADAR SENSORS

by

**Nuno Paulino, João Goes and Adolfo Steiger Garção**

*Universidade Nova de Lisboa, Fac. Ciencias e Tecnologia, Depto. Engenharia Electronica, Quinta da Torre, 2825-114 Caparica, Portugal*

Springer

Nuno Paulino
Universidade Nova de Lisboa
Fac. Ciencias e Tecnologia
Depto. Engenharia Electronica
Quinta da Torre
2825-114 Caparica
Portugal
nunop@uninova.pt

João Goes
Universidade Nova de Lisboa
Fac. Ciencias e Tecnologia
Depto. Engenharia Electronica
Quinta da Torre
2825-114 Caparica
Portugal
jg@uninova.pt

Adolfo Steiger Garção
Universidade Nova de Lisboa
Fac. Ciencias e Tecnologia
Depto. Engenharia Electronica
Quinta da Torre
2825-114 Caparica
Portugal

ISBN 978-90-481-7872-8 e-ISBN 978-1-4020-8410-2



Printed on acid-free paper

9 8 7 6 5 4 3 2 1

springer.com

# List of Abbreviations and Acronyms

| | |
|---|---|
| ADC | Analog to Digital Converter |
| AGC | Automatic Gain Control. |
| CMOS | Complementary MOS |
| CP | Charge Pump |
| DAC | Digital to Analog Converter |
| DC | Direct Current |
| DLL | Delay Locked Loop |
| DSP | Digital Signal Processor |
| DTC | Digital to Time Converter |
| EIRP | Effective Isotropic Radiated Power |
| ESD | Electro-Static Discharge |
| FCC | Federal Communications Commission. |
| FFT | Fast Fourier Transform |
| FIB | Focused Ion Beam |
| GaAs | Gallium Arsenide |
| GBW | Gain Bandwidth product |
| GPR | Ground Penetrating Radar |
| IC | Integrated Circuit |
| IF | Intermediate Frequency |
| IFFT | Inverse FFT |
| JLCC | Ceramic Leaded Chip Carrier (J Lead) |
| LC | inductor-Capacitor circuit |
| LNA | Low Noise Amplifier |
| lsb | least significant bit |
| MiM | Metal isolator Metal capacitor |
| MOS | Metal Oxide Semiconductor |
| NF | Noise Figure |
| NMOS | N channel MOS transistor |
| PCB | Printed Circuit Board |
| PD | Phase Detector |
| PLL | Phase Locked Loop |
| PMOS | P channel MOS transistor |
| PRF | Pulse Repetition Frequency |
| PRI | Pulse Repetition Interval |
| PSD | Power Spectral Density |
| RADAR | RAdio Detection And Ranging |
| RC | Resistor-Capacitor circuit |
| RF | Radio Frequency |
| RLC | Resistor-inductor-Capacitor circuit |
| rms | root mean square |

| | |
|---|---|
| SAW | Surface Acoustic Wave |
| ΣΔ | Sigma Delta modulator |
| S&H | Sample and Hold |
| sinc | sin(x)/x function |
| SMD | Surface Mounted Device |
| SNR | Signal to Noise Ratio |
| TDR | Time Domain Reflectometry |
| UWB | Ultra Wide Band |
| VCDL | Voltage Controlled Delay Line |

# Contents

# Chapter 1
# Introduction

The objective of the work presented in this book is the study, design and implementation of a low-cost, low-power radar sensor, using CMOS technology. This radar sensor can have several applications, including:

- *Proximity sensor*: being able to detect any intruders that enter a safety perimeter, defined around the sensor.
- *Motion detector*: being able to detect motion of objects around the sensor.
- *Ground penetrating radar*: Since the radar sensor will use signals with a large bandwidth at low frequency, it will be capable of detecting artifacts buried at shallow depth in some types of soil.
- *2D/3D imaging*: by combining the outputs of several radar sensors and using signal processing techniques it is possible to obtain a 2D/3D image.

The operation principle of a radar system is very simple: an electromagnetic signal (pulse) is radiated, this signal travels at the speed of light (assuming that the signal propagates in space), when the signal encounters an object part of its energy is reradiated therefore creating an echo, the echo signal travels back to the point of radiation of the original signal at the speed of light. This signal can be detected and the time interval between the transmission of the signal and arrival of the echo can be measured. The capability of the radar to distinguish between two targets close to each other is dependent on the relation between the pulse width and the time it takes for the light to travel the distance between the two targets. If the pulse width is larger than this time, it is very difficult to distinguish the echoes from each target. In order to separate two targets placed close together it is important to use a small pulse width. The bandwidth of signal with a given time duration (pulse) is approximately equal to the inverse of the signal time duration. If a pulse signal is used to modulate a sine-wave carrier, the resulting signal will have approximately the same bandwidth as the pulse signal centered around the carrier frequency. So, if traditional narrow band signals are used, the carrier frequency needs to be very high, in order to obtain a relatively low fractional bandwidth. This high frequency would result in the need for expensive RF technologies such as GaAs. But if the carrier frequency is reduced, eventually using no carrier signal just the pulse signal, the resulting signal will still have the capability of resolving two targets placed close to each other. As an example if it is desired to distinguish between targets separated by a distance of 30 cm it is necessary to use a pulse width smaller than 1 ns. If this pulse is used to modulate a sine-wave signal, the carrier frequency might be as high as 20 GHz, but if the pulse is used without a carrier signal, the resulting signal will have most of its power below 2 GHz. This frequency range is within the reach of modern CMOS technologies (0.35 μm, 0.18 μm or better), which are less expensive than more exotic technologies.



N. Paulino et al., *Low Power UWB CMOS Radar Sensors*, 1–3.

A signal which has a very large bandwidth compared to its center frequency value is known as an UWB signal and it has been know and utilized since the end of the nineteenth century. The interest in these types of signals has increased in the last years due to the potential benefits of imaging and communications systems based on UWB signals. Although initially, the use of UWB signals was prohibited because these signals would interfere with narrow band RF signals, recently it was realized that if the power of the UWB signal is low, the interference is limited and UWB signals can share the same electromagnetic spectrum as narrow band RF signals. This allows using the same electromagnetic spectrum (which a scarce resource) for different applications. In 2002 the FCC recognized this and allowed to use UWB signals in a variety of applications, as long as the PSD of the signals was limited to certain levels. UWB systems radiate signals with very low power level, to reduce the interference with narrow band systems. This means that the range of operation for UWB systems will be short; however by exploring the high bandwidth of UWB signals it is possible to obtain high data rate digital communications at short ranges or creating high-resolution radar images.

This book is divided in 6 chapters; a brief overview of the following chapters is given next.

The second chapter introduces UWB signals and systems. The utilization of UWB systems, in the last one hundred years, is described in a brief history of UWB signals and systems, showing that this type of signals were used in the beginning of the radio history, but were abandoned in favor of narrow band RF signals. During a long time UWB signals survived in niche applications. However, in the last 15 years, the evolution of the CMOS technology enabled a reduction of the cost of UWB systems and, at the same time, enabled an expansion of their functionality. This led to a renovated interest in UWB systems, both for communications and for imaging applications. This chapter shows some examples of UWB signals and characterizes the time domain and frequency domain properties of these signals. It is shown, that baseband UWB signals can be easily generated using circuits built in a CMOS technology, which is less expensive than more exotic technologies such as GaAs or Silicon-Germanium. The second chapter ends with a study of the interaction between UWB signals and the antennas used to transmit and receive them. In this analysis, a methodology that allows to calculate the shape of the received UWB signal in an antenna link is developed. Using this methodology, it is shown that the optimal input impedance of an UWB receiver is not necessarily the same as the optimal input impedance for a narrow band signal. There is also an analysis about the best type of antenna for UWB signals and how each type of antenna affects the PSD of the radiated signal.

The third chapter describes the operation of a radar system. The differences and advantages of using UWB signals in a radar system, over traditional narrow band signals, are discussed. The radar equation, usually defined for narrow band signals, is redefined for UWB signals. This new radar equation is used to analyze the echo signals from targets with basic shapes, resulting in an estimative of the echo signal amplitude as a function of the target distance. The architecture of an UWB radar based in the concept of sub-sampling is defined and the constituting blocks are described. The remaining of the chapter is concentrated in analyzing the receiver channel of the radar system. This analysis is centered around the noise power level in the receive path. From this analysis a new receiver circuit for baseband UWB signals is proposed. This new circuit is an averaging switched-integrator circuit. This circuit is described in detail and a design procedure is shown.

The fourth chapter deals with the problem of generating a clock signal with a programmable delay. This clock is necessary in the radar system to define the sampling instant in the receiver channel. The delay in this clock signal is relative to the transmit clock signal and is used to determine the target distance. A new digitally programmable delay architecture that can have a large programming linearity is proposed. This architecture is based on a digital $\Sigma\Delta$ modulator that controls a 1-bit digital-to-time converter, whose output is filtered by a delay locked-loop, thus producing a delayed clock signal. This architecture is analyzed and designed, at high level, to meet the required specifications for the radar system. A high level model is used to simulate the behavior of this circuit and validate the design. The electronic sub-blocks necessary to build this circuit are described, analyzed and a design methodology for each sub-block is derived. These circuits are implemented using differential clock signals in order to reduce the noise level in the radar system.

The fifth chapter describes the experimental CMOS prototype IC that was developed to evaluate the performance of the different sub-blocks of the radar system. The first part of this chapter is dedicated to the description of the prototype IC and to the design of the short pulse generator circuit and the amplifier circuit needed in the receive path. The second part of the chapter describes the experimental evaluation of the circuit and presents the measured results for each of the radar system sub-blocks. The experimental evaluation procedures used during the testing of the prototype IC, are also described.

Finally, the sixth chapter describes an overview of the research presented in this book, draws some conclusions about obtained results and presents some suggestions for future work in this area.

# Chapter 2
# UWB Signals and Systems

**Abstract** this chapter introduces UWB signals and systems. The utilization of UWB systems, in the last one hundred years, is described in a brief history of UWB signals and systems. This chapter shows some examples of UWB signals and characterizes the time domain and frequency domain properties of these signals. It is shown, that baseband UWB signals can be easily generated using circuits built in a CMOS technology. The chapter ends with a study of the interaction between UWB signals and the antennas used to transmit and receive them. In this analysis, a methodology that allows to calculate the shape of the received UWB signal in an antenna link is developed. Using this methodology, it is show that the optimal input impedance of an UWB receiver is not necessarily the same as the optimal input impedance for a narrow band signal. There is also an analysis about the best type of antenna for UWB signals and how each type of antenna affected the PSD of the radiated signal.

## 2.1 A Brief History of UWB Systems and their Applications

The first intentional emission of electromagnetic radiation was based in the generation of electrical signals with short duration [1]. In his pioneering experiments, Heinrich Hertz [2] invented a RF oscillator consisting of a resonant circuit (a balanced half wave length dipole, capacitively loaded by large metal spheres) and a fast acting switch (a spark gap in the center of the dipole). This arrangement was operated by applying a high voltage DC pulse to the dipole, this high voltage DC pulse was generated by the secondary of a transformer where a large DC current ($i_1$) was interrupted in the primary. A simplified representation of this circuit is shown in Fig. 2.1.

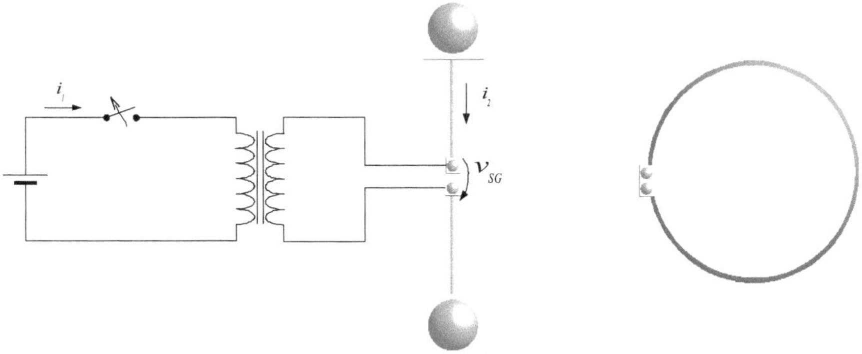


**Fig. 2.1** Simplified representations of the Hertz RF generator and receiver



N. Paulino et al., *Low Power UWB CMOS Radar Sensors*, 5–47.

The capacitance of the dipole is charged by the high voltage pulse until the voltage across the spark gap ($v_{SG}$) is enough to cause a spark. The spark arc has a low resistance that effectively connects the two parts of the dipole causing a large abrupt current ($i_2$) to flow between the two charged spheres. This current causes a magnetic field that reaches its maximum value when the electrical field between the two spheres is zero. As the magnetic field collapses, the electrical field increases (with an opposite polarity as before) because the current continues to flow charging the spheres, when the current reaches zero, the energy is once again stored in the electrical field and the cycle will repeat itself. The charge will thus flow back and forth between the spheres, as the energy alternates between the magnetic and the electrical field, except that approximately 15% of the energy is radiated as RF energy in each half cycle. The dipole produces about five half-cycles of RF energy for each high voltage pulse applied. The typical output wave of this oscillator is a short duration damped sine-wave and the resulting RF frequency spectrum is very wide.

In his experiments Hertz used his oscillator apparatus to generate electromagnetic waves with frequencies[1] of 50 MHz, 100 MHz and 430 MHZ; he used such high frequencies in order to obtain a wave-length with an acceptable small value, necessary to run his experiments inside his laboratory. He detected the electromagnetic waves using a very small spark gap connected to a resonant circuit consisting of a loop with half wave-length (as shown in Fig. 2.1). A small spark would indicate the presence of an electromagnetic field; this setup is not very sensitive and requires almost complete darkness to detect the small spark. Nevertheless, he was able to measure the wave length by determining the position of the peaks and nulls of the radiated RF signal.

Latter Marconi used and perfected Hertz experiments in order to obtain a wireless telegraphy system. His systems used variable capacitors to provide tuning to both receivers and transmitters, in order to select different frequencies of operation [3]. The transmitters were still based in spark plugs and therefore a considerate amount of RF energy was spread outside the band of interest, resulting in inefficient operation. The receivers were based on coherer tubes, a glass tube filled with metallic powder, whose resistance would drop approximately an order of magnitude when submitted to an electrical discharge. The sensitivity of this detector is very low. These systems were suited for communication using only Morse code. To compensate the large inefficiencies of the transmitters and of the receivers, the transmitters needed to radiate a large amount of power (over 10 kW). The result was, as the number of spark plug transmitters increased, the interference between them prevented clear communication between a given transmitter and a given receiver. This led to the invention and development of continuous wave oscillators, where the output signal is a continuous sine wave with a desired frequency value, resulting in a reduction of the power transmitted outside the desired frequency band and therefore a reduction in the interference between different transmitters.

[1] This is the fundamental frequency of the damped sine-wave pulse which is very rich in harmonics.

The sensitivity of the receiver circuits was improved, with the discovery of the first asymmetrical electrical junctions, such as the point contact crystal detector, which also improved the range and quality of the communication channel. The invention of the vacuum tube by DeForest in 1906 led to new circuits [4], [5], such as the superheterodyne receiver (invented by Edwin Armstrong around 1917) [6], which has an excellent frequency discrimination capability and sensitivity (this radio architecture is still the base for most of the radios used today). As a result, gradually the spark plug transmitters were abandoned and finally prohibited in 1927, since their large spectrum usage would interfere with narrow band RF systems.

Interest in radio systems based on short pulses never disappeared, during the 1940's several patents were issued for systems using short pulses such as the patent for a "Random Impulse System" issued in 1942 (US patent 2671896), but no practical applications for these systems existed. The practical use of non-sinusoidal RF signals reappeared in the 1960's with the objective of analyzing the properties of distributed microwave networks, first as an analytical tool but latter, with the development of synthetic generators and sampling oscilloscopes, these signals, called baseband pulses or carrierless pulses, were used to experimentally determine the intrinsic properties of materials and distributed networks [7], [8], by analyzing the reflections created by the input pulses, this type of analysis is know as Time Domain Reflectometry (TDR).

The synthetic generator consisted of a distributed pulse forming network, where a step function (created by a tunnel diode source) with sub-nanosecond rising time was applied. The pulse forming network had open and short circuit stubs located at precise distances in order to transform the incident step to a rectangular pulse and by appropriately inverting, recombining and filtering these pulses, into an output consisting of several RF cycles [8], [9].

The sampling oscilloscope uses equivalent time sampling to acquire signals whose bandwidth exceed half of the maximum sample rate of the oscilloscope [10]. The input signal is applied directly to a sampling bridge and is sampled before any attenuation or amplification is performed. The attenuator or amplifier circuits are placed after the signal has been converted to a lower frequency by the sampling circuit, and can have a lower bandwidth than the input signal. Therefore the sampling oscilloscope can have an input bandwidth much larger than the sample rate. The equivalent time sampling takes advantage of the fact that most man made events can be repetitive. Equivalent time sampling constructs an image of a fast repetitive signal by sampling a point of the input signal in each repetition. Each time, a different point is sampled and after a certain time it is possible to reconstruct the shape of the fast input signal. This principle, known as sub-sampling, is illustrated in Fig. 2.2 This process allows a sampling oscilloscope to display signals whose frequency components are much higher than the oscilloscope sample rate (but inferior to the sampling bridge input bandwidth).

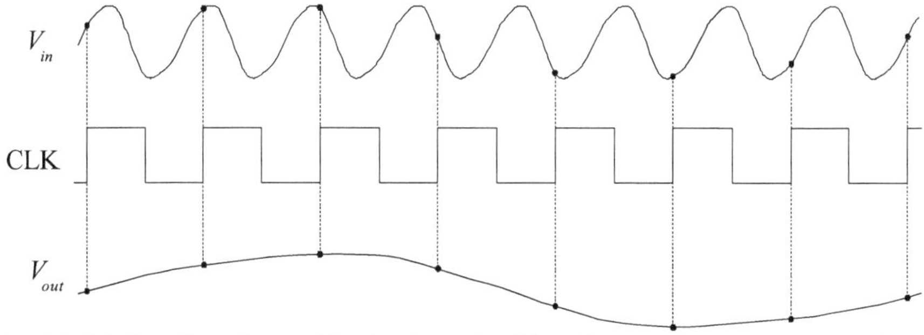


**Fig. 2.2** Sub-Sampling of a repetitive fast input signal in order to produce a slow output signal with the same shape as the fast input signal

Using pulses with very short duration it is possible to measure the impulse response (through a sampling oscilloscope) of a distributed microwave network and then calculate the frequency response using the Fourier transform. The maximum and minimum frequency values of the analysis are determined by the pulse width (the shorter the pulse the higher the maximum frequency value of the analysis) and by the duration of the observation time (the longer the observation time the lower the minimum frequency value of the analysis). Of course that increasing the maximum and the minimum frequency of the analysis requires both a fast sampling frequency in the oscilloscope and a large number of sampled points of the impulse response. This procedure reduced the time needed to measure the frequency domain properties of radar stealthy materials by a factor of 100 [8]. The traditional method consists in applying a sine-wave signal, measuring the S parameters and then repeating the procedure for the different frequencies of interest. This is the method of choice for narrow band systems. But for wide band systems, such as today's digital systems (Rambus, USB 2.0, Firewire and others), it is difficult to interpret the measured frequency response and understand where the discontinuities in the transmission lines are physically located. TDR provides a more intuitive and direct look into the signal integrity in digital systems [11], [12]. By using a step generator to generate a fast edge into the transmission line and using a sampling oscilloscope to monitor the reflected waves at different points in the line it is possible to locate the line discontinuities and understand the nature (resistive, inductive or capacitive) of each discontinuity. It is also possible to understand whether losses in a transmission system (a circuit board trace, a cable, a connector and so on) are series losses or shunt losses.

TDR has also been used to locate various defects in electrical power distribution cables (such as: voids, shield protrusions, contaminants, etc.) by locating partial discharge events in the cable [13], [14]. The cable under test, terminated with an open end, is connected to an excitation voltage source at the other end. This voltage is gradually increased until a partial discharge (PD) event occurs at a certain point in the cable; the pulses generated by the PD event travel in the cable and are reflected by the ends of the cable until they disappear because of the cable attenuation. By measuring the time between the arrivals of the various pulses at one end of the cable it is possible to determine where the defects are located.

Another area of application for TDR techniques is liquid level measurement in tanks [15], [16]. The level of the liquid (or liquids) in a tank can be determined using a probe line that acts as an open transmission line for the TDR pulse signal. The tank contents that surround the probe line become the transmission line isolating dielectric. At each fluid interface, a change in the dielectric constant of the fluid surrounding the probe results in a change in the electrical impedance of the line that reflects a portion of the input pulse amplitude to the TDR receiver. The Pulse reflection amplitude and timing information can be used to accurately determine the fluids levels and the location of each fluid interface within the tank (if more than one type of fluid is present in the tank).

In the 1970's it became apparent that the sub-nanosecond pulse technology could be employed in a variety of proximal sensing applications [8], [17], this is essentially extending the time domain reflectometry principles to work in free space without a transmission line, this is also know as free-space time-domain reflectometry. Since in those systems, the objective is simply to determine the presence or absence of a return signal from a possible target, the return sensing system (receiver) can be designed to determine the threshold (location) information only, leading to significant reductions in cost and complexity of the required electronics for target signature analysis. The receiver senses, over a narrow time window, when the input signal exceeds a given threshold value, indicating the presence of a target and determines the time delay between this event and the initial transmission of the sub-nanosecond pulse. The distance to the target is directly related to this time delay. This type of systems can be used to establish security “bubbles”, where an alarm is triggered if something enters the protected area. In [8] two types of receivers are described. The first, called single-hit receiver is based on a tunnel diode that is triggered by the input very-fast low energy impulse. This receiver is depicted in Fig. 2.3.

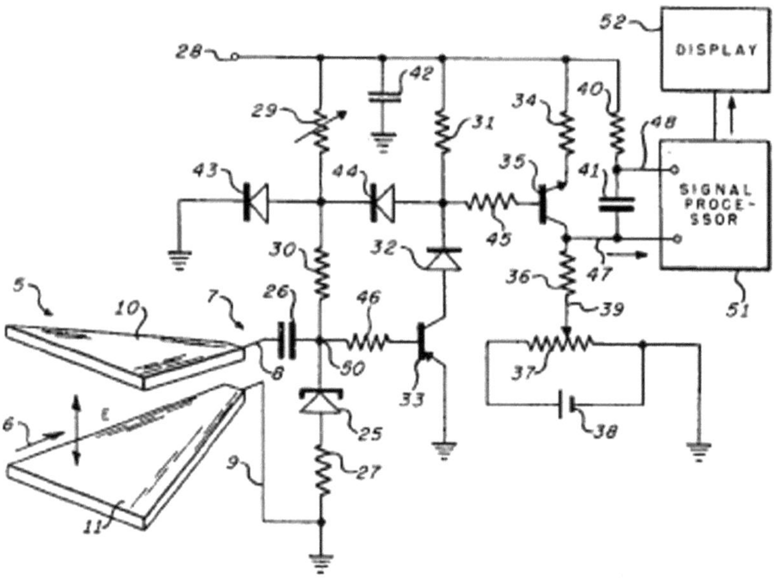


**Fig. 2.3** Single-Hit receiver using a tunnel diode. Illustration from the US patent 3662316 "Short Base-Band Receiver" issued to Kenneth Robins

The operation of this circuit is as follows:[2] the tunnel diode (25) is biased close to its break down point, by adjusting its bias current using variable resistor (29). In this situation the voltage in node (50) is not enough to turn transistor (33) on . When a small amplitude (around 100 mV) short pulse (around 0.5 ns) is received by the antenna (10) and applied to the tunnel diode, the diode will instantly switch from its low voltage, positive resistance state through its unstable negative resistance state to its high voltage positive resistance state, causing the voltage at node (28) to drop. This will turn on transistor (33), causing a current to flow from the collector of this transistor through the diode (32) and the resistor (31). This current causes an increase in base voltage of the avalanche transistor (35) until the transistor (35) is turned on abruptly, essentially shorting the emitter and collector, causing the base voltage of this transistor to change very fast to a positive value. This voltage is coupled through diode (44) and resistor (30) to the tunnel diode causing the diode to cut off (the maximum voltage value at this node is clamped by diode (43)), which in turn also turns off transistor (33). The output voltage (48) suffers a sharp negative edge when transistor (35) turns on and then a slow decay, as capacitor (41) is discharged through resistor (40) and (34) and transistor (35). This indicates that a short pulse (hit) was detected by the receiver. A certain time after, transistor (35) turns off. The voltage at node (50) returns to its initial value and the

[2] Note that node (28) has a negative voltage compared to the ground.

circuit is ready to detect another fast pulse at the input. In conclusion this circuit works as a monostable circuit, that when a small amplitude short pulse is presented to its input, produces a large amplitude pulse with a long duration at the output. This output pulse can be used to measure the time between the transmitted pulse and the return pulse created by the target (echo) and therefore, determine the distance of the target. The clock signal that triggers the transmission of the short pulse can be applied to a transmission line with the necessary time delay for the pulse to travel to target and back to the receiver; this delayed clock can be used to sample the output of the receiver, to determine if a target is present at the predetermined distance. The sensitivity of this receiver is determined by the tunnel diode, if this device is biased closer to its turning point, the amplitude value of the input pulse necessary to trigger the diode is reduced. This means that the circuit threshold has been lowered and it will be more sensitive to any input pulses, but at the same time it will be more prone to trigger on unwanted signals, i.e. circuit noise.

The second type of receiver described in [8], called a constant false-alarm rate receiver, improves the sensibility of the single-hit receiver by processing several hits before signaling the detection of the target. The receiver is essentially the same, except that the bias current of the tunnel diode is adjusted by a feedback loop. The objective is to bias the diode as close as possible to the threshold of the negative resistance region, to increase the sensitivity of the receiver. The feedback loop is controlled by the number of hits detected when no signal is present. This is archived by radiating the pulse on the rising edge of a clock and sampling the output of the single-hit receiver on the falling edge of the clock. As the sensitivity of the receiver is increased the number of false hits increases, the feedback loop will adjust the bias current of the tunnel diode until the number of false hits is equal to 3 over 32 observation time instants. This means that the probability of a false hit is around $(1/16)$, the probability of 32 consecutive false hits would be equal to $(1/16)^{32}$ which is a very small number. The output of the single hit receiver is sampled when the clock is present and the number of hits obtained is added, a target is considered to be present when this sum exceeds 25, this will eliminate most of the false hits. A proximity radar for detecting incoming ammunitions, using this receiver, is described in [18].

In the early 1990's, McEwan [19] developed a low cost, low power impulse radar system, using discrete components mounted on a small PCB; this sensor is called Micropower Impulse Radar (MIR). This radar sensor uses impulses with a duration between 1 and 2 ns. These pulses are generated by applying a step signal to a step recovery diode. The radar transmitter emits these pulses with a random interval to reduce the probability of interference between different radar sensors. Each sensor has a random noise generator that modulates the period of the square wave that is used to generate the short pulses. A simple block diagram of the impulse transmitter is shown in Fig. 2.4

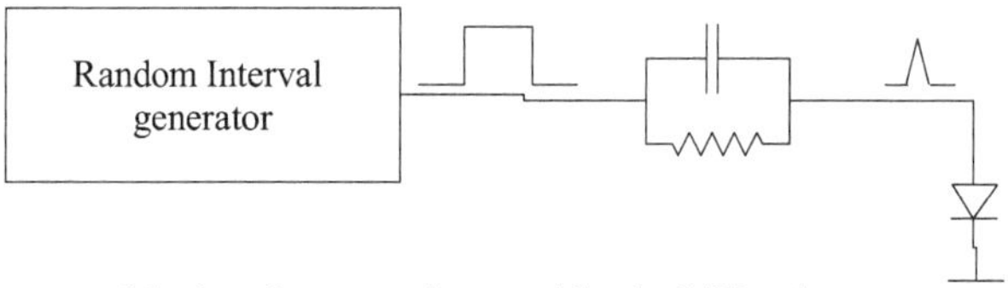


**Fig. 2.4** Block diagram of the impulse transmitter used by the MIR radar sensor

The short pulses are generated using a step recovery diode [20]. Normal diodes have a slow transition when the current is switched from forward to backward, the step diode is designed to have an abrupt transition when the switch occurs, thus producing a short pulse.

The pulse transmitter has nominal pulse repetition interval (PRI) typically equal to 1 μs, meaning that on average 1 million impulses per second are radiated. The receiver is constituted by two single ended integrating samplers. These circuits are constituted by a diode switch followed by an analog integrator. This integrator averages the input samples, if no signal is present the average output of the integrators will not change, if a signal is present the average output of the integrator will increase. This configuration integrates several impulses to improve the circuit sensitivity, taking advantage of the large number of available impulses. The input signal is sampled into an integrating capacitor when a strobe signal is applied. This strobe signal is generated from the random interval generator using a circuit equal to the one depicted in Fig. 2.4. The strobe signal determines the range at which a target might be located, thus a delay must be inserted between the random interval generator and the short pulse generator circuit. In a simple application, such as intrusion detection this delay can be constant and its value determines the value of the security perimeter.

Another area where UWB signals have been applied is the field of subsurface radar. The objective is to detect and locate buried artifacts and structures within the upper regions of the earth's surface [21], [22]. Any dielectric variation in the buried materials, produces a reflection from an incoming electromagnetic signal, this reflection can be used to determine the depth and type of buried material. Since the materials that constitute the earth structure introduce a high signal attenuation, the maximum operating depth of these radars is usually limited to a few meters (the signal attenuation ranges from 10 dB per meter to 200 dB per meter, depending on the material). The signal attenuation is dependent on the type of materials of the soil and, in general, it increases with the signal frequency. The maximum useful frequency is below 1 GHz, for example wet clay soil has an attenuation of 30 dB per meter for a frequency value of 100 MHz and an attenuation of 100 dB per meter for a frequency value of 1 GHz [21]. In most cases, subsurface radar systems are forced to work with signal frequencies below 1 GHz but, need to have a high-resolution to detect buried artifacts at shallow depths; therefore baseband pulses with a short duration that inherently have a large bandwidths are necessary. Typically, pulses with an amplitude of 100 V and a duration between 1 ns and 5 ns are used. The received signal is down-converted using a sampling circuit and digitized, so that complex signal processing algorithms can be applied.

Recently, UWB ground penetrating radars (GPR) have been used to detect buried mines [23], [24]. In reference [23] a GPR system using a 4 m wide array constituted by 16 UWB transmitters and 16 UWB receivers is used to build 2D and 3D images of the ground beneath the sensor. The transmitted pulses have a repetition frequency of 1 MHz and have their energy spread from 100 MHz to 3 GHz. The receivers are constituted by an RF gain stage followed by a sampler and an ADC. This type of GPR is capable of reliably detecting buried mines to a depth 50 cm.

All the previous applications show that systems using UWB signals can be as useful as narrow band systems and some of the applications are only possible using UWB systems. As a result, the use of UWB systems has been increasing over time. In the last ten years the interest in UWB systems has increased dramatically [25], this can be attributed to a simple reason: cost.

Narrow band RF systems use tuned circuits to select between different signals. These circuits are built using passive components such as inductors, capacitors or SAW filters. Until the cost of active components (transistors) started to drop exponentially (due to the constant improvement of integrated circuit technology), electronic designers minimized the number of active components in their designs to reduce cost. This is the reason why old RF systems have many passive tuned circuits that needed to be calibrated ("tuned") before the equipment leaves the factory. As the cost of integrated circuits dropped (following Moore's Law [26]), designers started to replace passive tuned circuits by equivalent circuits using transistors, such as active filters and switched-capacitor filters. After the cost of digital signal processors (DSP) become affordable, many systems started to implement filters on DSPs (this also requires analog-to-digital and digital-to-analog converters). These techniques also reduce the need for calibration. The cost of passive components does not change much, throughout the years; the values of a capacitance, or an inductance, or a resistance, are strongly dependent on the component physical dimensions and constituting materials; therefore it is difficult to reduce its cost. On the other hand, as the cost of integrated circuits reduces (and also their size), their performance has also been doubling every 18 months. This has lead to the development of electronic designs where there are less and less passive components and more advanced signal processing algorithms are used (this is even more dramatic if the ratio between the number of passive and active components is considered). A good example is the Armstrong superheterodyne receiver architecture (which is still the base for all modern radio receivers) where the last mixer is currently implemented in a DSP through the multiplication of digital signals [27]. Even the RF signals that are received are now modulated by digital signals instead of analog signals.

UWB systems do not require tunned circuits, the signals can be selected in the time domain. This makes UWB signals very amenable to digital processing techniques. These techniques can be used to improve the sensibility of UWB receivers lowering the required input RF power level, thus reducing the probability of interference with traditional narrow band systems (the main problem that plagued the first spark plug systems). Since UWB signals require a large bandwidth, it is required that the circuits processing UWB signals need to be very fast. In the past the cost of high speed circuits was prohibitive, but, as previously stated, the cost of integrated circuit technology is dropping and the performance (speed) is increasing over time. This makes the cost of UWB systems drop over time. In the last years, there is an increasing number of proponents for the use of UWB communication systems [28], [29], [30], [31]. The objective is to use UWB signals to create short communication links (up to 10 m) with high data-rate (over 100 Mbits/sec). The required power level for the UWB signals is very low and therefore the interference between these signals and traditional narrow-band RF systems is unlikely.

In 2002 the Federal Communications Commission (FCC) [32] recognized that: "UWB technology holds great promise for a vast array of new applications that we believe will provide significant benefits for public safety, businesses and consumers. With appropriate technical standards, UWB devices can operate using the spectrum occupied by existing radio services without causing interference, thereby permitting scarce spectrum resources to be used more efficiently". The reasoning is that since UWB devices operate using very narrow or short duration pulses that result in a very large transmission bandwidth, UWB devices can operate using spectrum occupied by existing narrow band radio services without causing interference, thereby permitting scarce spectrum resources to be used more efficiently.

UWB technology was defined as any wireless transmission device using signals whose fractional bandwidth ($B_{frac}$) is greater than 0.25 or occupy 1.5 GHz or more of spectrum. The formula proposed by the Commission for calculating fractional bandwidth is:

$$B_{frac} = \frac{2 \cdot (f_H - f_L)}{f_H + f_L} \tag{2.1}$$

where $f_H$ is the upper frequency of the –10 dB emission point and $f_L$ is the lower frequency of the –10 dB emission point. The center frequency of the transmission was defined as the average of the upper and lower –10 dB points:

$$f_c = \frac{f_H + f_L}{2} \tag{2.2}$$

Any electronic or electrical equipment always emits unwanted radiation that can interfere with narrow band radio systems. This is especially true for digital systems such as computers. Therefore every electronic and electrical device is required to limit the power of these unwanted emissions below a certain level. If UWB systems operate with power levels below the unwanted emissions level, there should not be any interference problems with other radio systems. The FCC defined three types of UWB systems: imaging systems, vehicular radar systems and communications and measurement systems. The applications and authorized operating frequency bands for these systems are defined as follows:

1. Imaging systems include Ground Penetrating Radars (GPRs), wall imaging systems, through-wall imaging systems, surveillance systems and medical imaging devices. The utilization of this type of systems is restricted to certain users (law enforcement, fire and rescue organizations, scientific research institutions, commercial mining companies and construction companies).
   1.1. GPRs must be operated with their –10 dB bandwidth below 960 MHz or in the frequency band 3.1–10.6 GHz. GPRs operate only when in contact with, or within close proximity of, the ground for the purpose of detecting or obtaining the images of buried objects. The energy from the GPR is intentionally directed down into the ground for this purpose.
   1.2. Wall-imaging systems are designed to detect the location of objects contained within a "wall," such as a concrete structure, the side of a bridge, or the wall of a mine. Wall imaging systems must be operated with their –10 dB bandwidth below 960 MHz or in the frequency band 3.1–10.6 GHz.
   1.3. Through-wall Imaging Systems must be operated with their –10 dB bandwidth below 960 MHz or in the frequency band 1.99–10.6 GHz. Through-wall imaging systems detect the location or movement of persons or objects that are located on the other side of a structure such as a wall.
   1.4. Surveillance Systems will be permitted to operate with their –10 dB bandwidth in the frequency band 1.99–10.6 GHz. Surveillance systems operate as "security fences" by establishing a stationary RF perimeter field and detecting the intrusion of persons or objects in that field.
   1.5. Medical Systems must be operated with their –10 dB bandwidth in the frequency band 3.1–10.6 GHz. A medical imaging system is used to detect the location or movement of objects within the body of a person or animal.
2. Vehicular radar systems using directional antennas on terrestrial transportation vehicles provided the center frequency of the emission and the frequency at which the highest radiated emission occurs are greater than 24.075 GHz. The –10 dB bandwidth must be between 22 and 29 GHz. These devices are able to detect the location and movement of objects near a vehicle, enabling features such as near collision avoidance, improved airbag activation, and suspension systems that better respond to road conditions. Attenuation of the emissions below 24 GHz is required above the horizontal plane in order to protect space borne passive sensors operating in the 23.6–24.0 GHz band.
3. Communications and measurement systems, such as high-speed home and business networking devices as well as storage tank measurement. The devices must operate with their –10 dB bandwidth in the frequency band 3.1–10.6 GHz. The equipment must be designed to ensure that operation can only occur indoors or it must consist of hand held devices that may be employed for such activities as peer-to-peer operation. The limits on unwanted emissions are more stringent for hand held devices than they are for indoor-only systems.

The main reason for defining the above operating frequency bands for UWB systems was due to the concern of the possible interference to the GPS band at 1559–1610 MHz and the planned L5 frequency in the 960–1215 MHz band.

The specified average emission limits of the different UWB systems for the different frequency bands are given in Table 2.1. These limits are defined in terms of dBm EIRP (equivalent isotropically radiated power) as measured with a one megahertz resolution bandwidth.

**Table 2.1** Average emission limits (in dBm/MHz EIRP) for UWB systems specified by the FCC

| Frequency (MHz) | Indoor communication systems | Hand held communication systems | Low frequency imaging systems | Medium frequency imaging systems | High frequency imaging systems | Vehicular radar systems |
|---|---|---|---|---|---|---|
| <960[3] | –41.3 | –41.3 | –41.3 | –41.3 | –41.3 | –41.3 |
| 960–1610 | –75.3 | –75.3 | –65.3 | –53.3 | –65.3 | –75.3 |
| 1610–1900 | –53.3 | –63.3 | –53.3 | –53.3 | –53.3 | –61.3 |
| 1900–1990 | –53.3 | –61.3 | –51.3 | –51.3 | –51.3 | –61.3 |
| 1990–3100 | –51.3 | –61.3 | –51.3 | –41.3 | –51.3 | –61.3 |
| 3100–10600 | –41.3 | –41.3 | –51.3 | –41.3 | –41.3 | –61.3 |
| 10600–22000 | –51.3 | –51.3 | –51.3 | –51.3 | –51.3 | –61.3 |
| 22000–29000 | –51.3 | –51.3 | –51.3 | –51.3 | –51.3 | –41.3 |
| 29000–31000 | –51.3 | –51.3 | –51.3 | –51.3 | –51.3 | –51.3 |
| >31000 | –51.3 | –51.3 | –51.3 | –51.3 | –51.3 | –61.3 |

Note that –41.3 dBm/MHz corresponds to 74 nW/MHz. The total power allowed for transmission of an UWB system depends on the frequency band used by that system, the maximum power allowed for transmission (in a given frequency band) is obtained integrating the value of the EIRP from Table 2.1 over the frequency band. As an example, for an UWB system that uses the frequency band from 0 Hz to 960 MHz, the maximum power allowed is 71.2 μW and an UWB system that uses the frequency band from 3.1 GHz to 10.6 GHz has a maximum allowed power equal to 533.7 μW. These values correspond to the total power emitted and they

[3] For frequencies below 960 MHz all UWB devices are permitted to emit at or below the part 15.209 limit of –41.3 dBm/MHz EIRP which is equivalent to an electrical field of 500 mV/m/MHz at 3 m distance.

assume that the UWB signal has a constant power spectral density value over the frequency band and has no power outside the frequency band. Normally this is not the case, the power spectral density of UWB is not constant and it is not easy to absolutely confine it to a given frequency band. These factors cause that the value of practical maximum power transmitted by an UWB system is below the previous values.

## 2.2 Spectrum and Power of UWB Signals

There are several signals that can be classified as UWB signals; these are typically constituted by a repetitive sequence of short pulses with a certain repetition frequency (PRF). The shape of the pulse, PRF value and the modulation (if any) determines the power spectral density (PSD) of the signal. The electromagnetic spectrum occupied by a transmitted UWB signal is determined by the PSD of the signal and by the transfer function of the antenna used to radiate the signal.
The power spectral of a periodic signal $v(t)$, $S_v(f)$, can be calculated using the Fourier series of the signals as follows [33]:

$$S_v(f) = \sum_{n=-\infty}^{+\infty} \left| c_n(n \cdot f_0)^2 \right| \cdot \delta(f - n \cdot f) \tag{2.3}$$

where $f_0$ is the PRF and $c_n$ are the Fourier series coefficients given by:

$$c_n = \frac{1}{T_0} \int_{T_0} v(t) \cdot e^{-j2\pi f_0 t} dt \tag{2.4}$$

The PSD can be calculated by solving the integral (2.4) [34], [35], which can be difficult depending on the shape of the pulse and modulation, or it can be calculated numerically applying the FFT algorithm to compute the discrete time Fourier Transform of a sampled version of the UWB signal [33], [36]. Both of these approaches are valid, the PSD presented next are calculated using the FFT. Note that both of these methods result in an approximated result, since, for most pulse shapes the PSD is constituted by an infinite number of spectrum lines and the graphic representation of their PSD is always truncated. The power of the signals will be calculated considering that the signals are voltage signals, normalized to 1 V amplitude ($A$) and the PSD will be presented using dBV (dB to 1 V *rms*) because in order to calculate the power level in dBm, it would be necessary to know the value of the load resistance where the signal is applied. If the resistance value is known ($R_L$), it is easy to convert the dBV value to dBm, considering that:

$$dBm(P_S) = 10 \times \log\left(\frac{V_S^2}{R_L \times 1mW}\right) = 20 \times \log\left(\frac{V_S}{1\,V_{rms}}\right) + 10 \times \log\left(\frac{(1\,V_{rms})^2}{R_L \times 1mW}\right)$$
$$= dBV(V_S) + 30dB - 10 \times \log\left(\frac{R_L}{1\,\Omega}\right) \tag{2.5}$$

The first signal to analyze is a rectangular pulse with a short duration. This signal is essentially a square-wave signal with a very low duty-cycle. The power of this signal is given by:

$$P = \frac{1}{T_{PR}} \cdot \int_{T_{PR}} v(t)^2 dt = A^2 \cdot \frac{\tau}{T_{PR}} \tag{2.6}$$

where $A$ is the signal amplitude, $T_{PR}$ the pulse repetition period and $\tau$ the duration of the pulse.

The spectrum of this signal is constituted by discrete lines spaced apart in the frequency by the value of the pulse repetition frequency, with an amplitude value modulated by the *sinc* function. The zeros of the *sinc* function are located at frequencies multiple of $1/\tau$. This signal and its spectrum are depicted in Fig. 2.5.

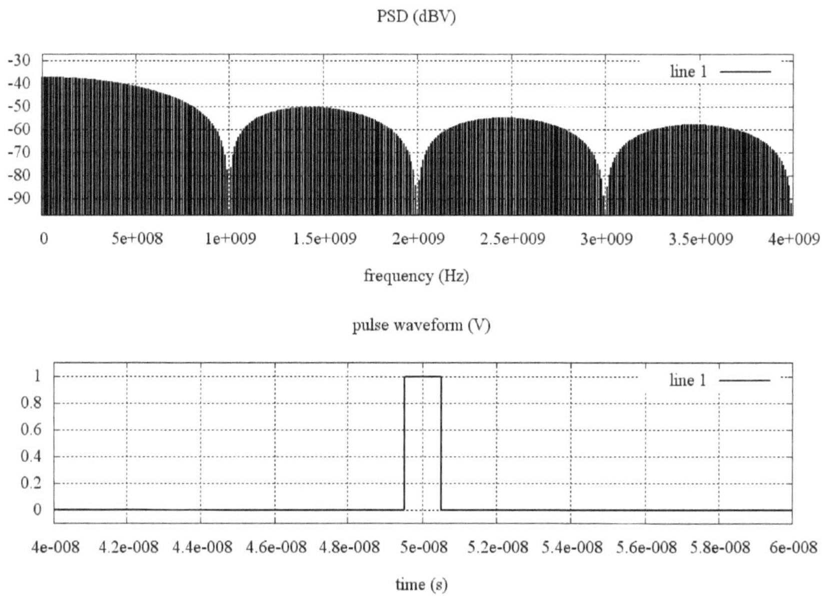


**Fig. 2.5** Power spectral density of a rectangular pulse with $A$=1 V, $\tau$=1 ns and PRF=10 MHz

The PSD of a rectangular signal has fractional bandwidth $B_{frac} = 2$ (all signals that have a DC component have a fractional bandwidth equal to 2 because $f_L$=0). If the PRF value is increased, while maintaining the same value of $\tau$, the signal power increases (expression 2.6) and the PSD discrete lines spacing and level also increases (expression 2.4).

The maximum PSD value is –36.3 dBV, assuming that this signal is applied to an antenna with a radiation resistance equal to 50 Ω and a maximum gain equal to 0 dBi, the maximum EIRP value would be equal to (using 2.5) –23.6 dBm and the total power of the signal is 200 μW. These values exceed the maximum allowed levels from Table 2.1 but, in reality, the antenna would act as filter that attenuates the PSD values resulting in a lower EIRP value as it will be shown in the next section. This shows that a possible UWB transmitter circuit can use a power supply voltage as low as 1.8 V and even assuming a radiation efficiency as low as 1% (due to parasitic and mismatch losses), the transmitter circuit would only need to dissipate 20 mW to transmit an UWB signal with the maximum power level allowed.

The level of the PSD is relatively high because the power is concentrated into discrete frequencies (multiples of the PRF) with no power in between. The level of the PSD can be lower if the value of the PRF is not fixed. This can be achieved by modulating the position of the pulse by a random signal, resulting in spreading the power of the signal over a larger frequency band, making the PSD level lower and therefore allowing for a larger total power level to be radiated. To illustrate this principle the estimation of the PSD of a signal where the position of the rectangular pulse changes randomly with a standard deviation of 1 ns is shown in Fig. 2.6, note that although the signal power is equal to the signal in Fig. 2.5, the PSD level is lower. In a communications link, this modulation can be the information that is being transmitted (assuming that the information content is random).

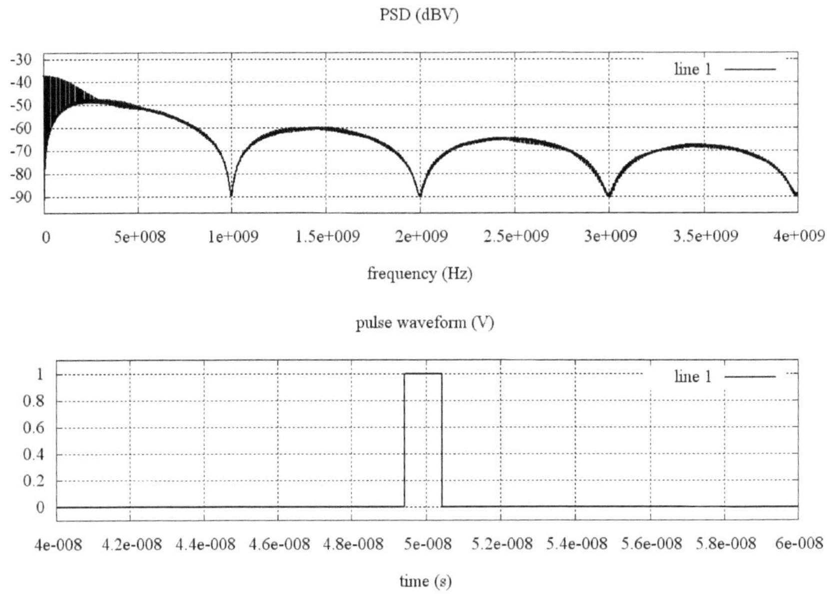


**Fig. 2.6** Estimation of the power spectral density of a rectangular pulse with $A$=1 V, $\tau$=1 ns, PRF=10 MHz and random dither with $\sigma$=1 ns

To illustrate the difference between narrow band signals and UWB signals, the next signal is a sine-wave whose amplitude is modulated by a square wave with a small duty-cycle. The sine-wave frequency value is 3 GHz, the value of $\tau$ (the active part of the square wave) is equal to 10.6 ns and the PRF is equal to 11.79 MHz. The PSD of the resulting signal corresponds to the PSD of a rectangular wave translated in frequency to the sine wave frequency. The bandwidth of this signal is determined by the bandwidth of the square wave signal, the smaller the duration of the rectangular pulse the larger the resulting bandwidth. The PSD of this signal is depicted in Fig. 2.7 together with the signal. The bandwidth of the signal is 100.6 MHz, which is fairly large, but the fractional bandwidth is only 0.039, indicating that most of the signal power is concentrated around the sine-wave frequency. The power of this signal is given by:

$$P = \frac{1}{T_{PR}} \cdot \int_{T_{PR}} v(t)^2 dt = \frac{A^2}{2} \cdot \frac{\tau}{T_{PR}} \tag{2.7}$$

the power is, as expected, the power of a sine-wave multiplied by the duty-cycle of the square-wave. It should be noted that the peak power (corresponding to the sine-wave power) can be much larger than the signal power.

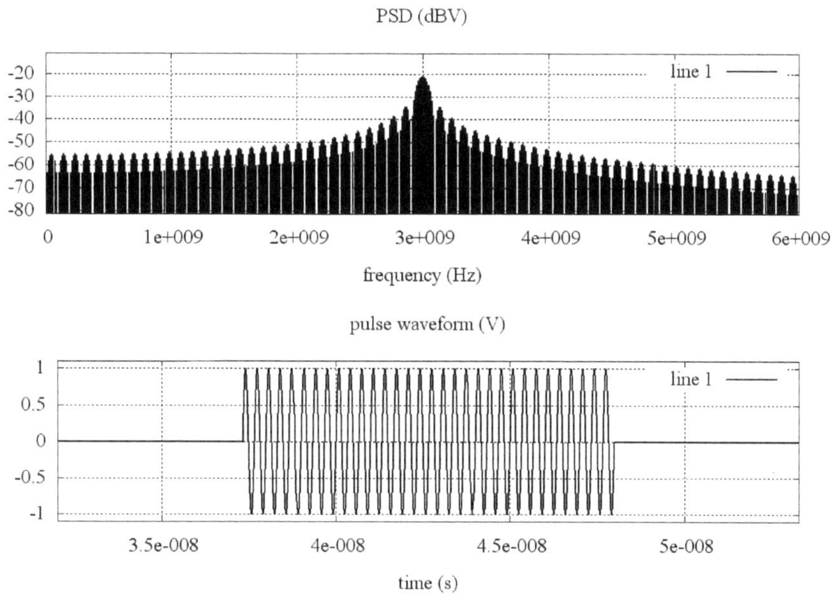


**Fig. 2.7** Power spectral density of a pulsed sine-wave signal with a center frequency of 3 GHz $\tau$=10.6 ns and PRF=11.79 MHz

If the duty-cycle of the square-wave is reduced, the bandwidth of the resulting signal increases. If the active time $\tau$ is reduced to 1.33 ns (while maintaining the sine wave frequency value) the resulting UWB signal has a fractional bandwidth equal to 0.365 and a bandwidth of 1.1 GHz. This signal, together with its PSD is depicted in Fig. 2.8.

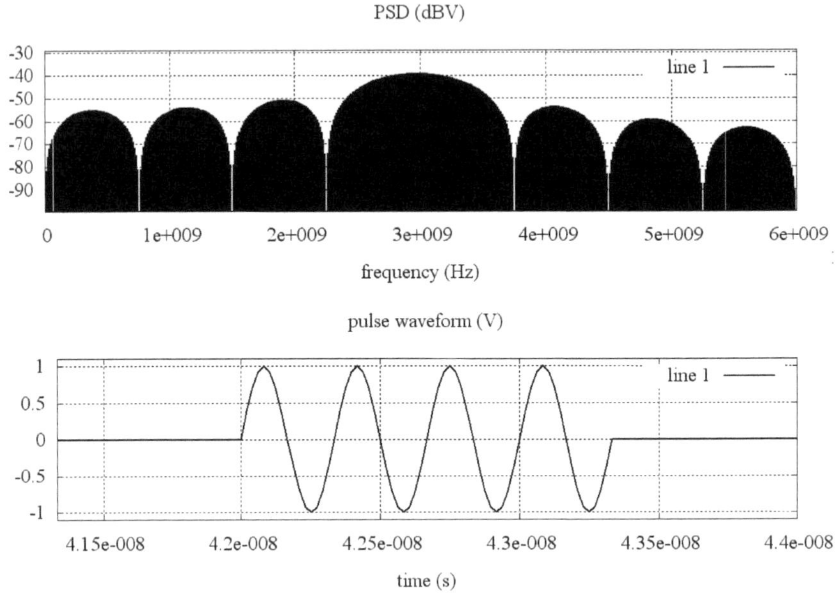


**Fig. 2.8** Power spectral density of a pulsed sine-wave signal with a center frequency of 3 GHz $\tau$=1.3 ns and PRF=11.79 MHz

The signal depicted in Fig. 2.8 has a PSD more spread over the frequency when compared to the signal in Fig. 2.7. If the sine-wave frequency is reduced, while maintaining the value of $\tau$, the fractional bandwidth of the signal would increase.

The rectangular signal of Fig. 2.5, is called a baseband signal because it has no carrier. However, in practice it is not possible to create such a signal because it has zero rise and fall times. A family of baseband signals that can be created is based on the Gaussian mono-cycle and its derivatives. These signals are described by the following expression [37]:

$$v_n(t) = \frac{d^n}{d^n t}\left[ e^{-\left(\frac{t}{\sigma}\right)^2} \right] \tag{2.8}$$

where $n$ is 0, 1 or 2 and $\sigma$ determines the width of the pulse. The signal obtained making $n$=0 is the Gaussian pulse, depicted in Fig. 2.9, together with its PSD. The PSD is similar to the PSD of the rectangular pulse at low frequencies, but there is almost no power at higher frequencies. This signal has a fractional bandwidth of 2 and $f_H$=680 MHz.

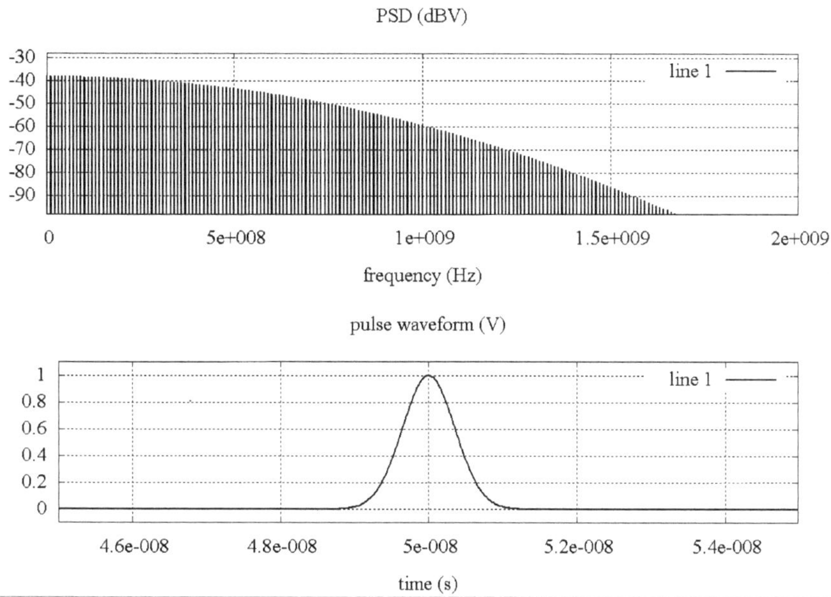


**Fig. 2.9** Power spectral density of a Gaussian pulse with $\sigma$=0.5 ns and PRF=10 MHz

The power of this signal is given by:

$$P = \frac{1}{T_{PR}} \cdot \int_{T_{PR}} v(t)^2 dt = A^2 \cdot \frac{\tau}{T_{PR}} \cdot \sqrt{2 \times \pi} \tag{2.9}$$

As in the case of the pulsed sine-wave it is possible to have a signal with low power but with a large peak power, corresponding to a large amplitude pulse, if $T_{PR} >> \sigma$.

To exemplify that it is easy to generate a signal similar to the Gaussian pulse, consider the circuit depicted in Fig. 2.10. This circuit uses CMOS circuits of a 0.35 μm technology. It generates a short pulse, with a duration determined by the delay of three inverters. The pulse is filtered by a simple RC circuit to obtain a "smooth" shape closer to the Gaussian pulse.

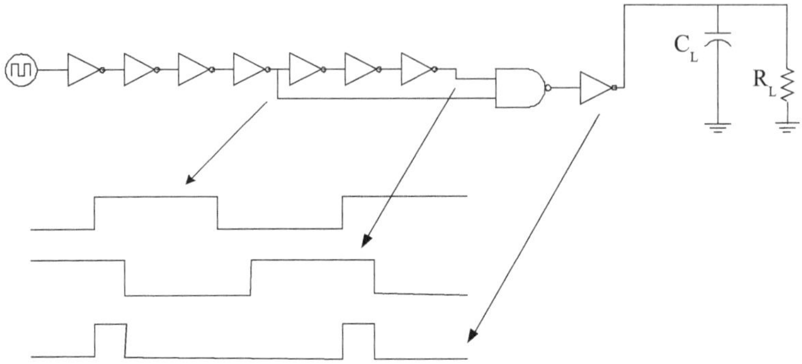


**Fig. 2.10** CMOS circuit for generating a short pulse. The logic circuit are CMOS and the minimum channel length of the transistors is 0.35 μm, the load capacitance value ($C_L$) is 3 pF and the load resistance ($R_L$) value is 180 Ω, this load together with the transistor's channel widths values determine the duration and shape of the pulse

The output simulated voltage of the circuit, together with its corresponding PSD is shown, in Fig. 2.11. The shape of this pulse is fairly similar to the Gaussian pulse shape and the power spectral density of this signal is also similar to the PSD of the Gaussian pulse. This signal has a fractional bandwidth equal to 2 (it has a DC component) and $f_H$=2.2 GHz. These results show that modern CMOS technologies can be used to build circuits capable of generating pulses with a duration inferior to 1 ns.

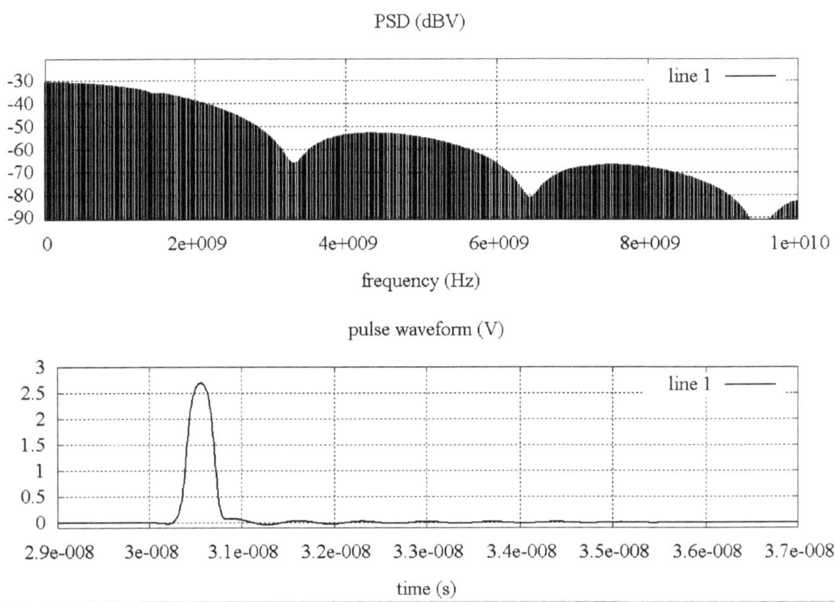


**Fig. 2.11** Pulse generated by the circuit in Fig. 2.10 and its corresponding power spectral density

The Gaussian pulse has a DC component that cannot be radiated. Considering that the antenna has a high-pass transfer function characteristic for low frequencies [37], [38], it is obvious that the transmitted signal will have a different shape from the input signal and therefore a different PSD. The PSD of the transmitted pulse will have less power at lower frequencies than the input signal; this makes the Gaussian pulse an inefficient signal to radiate, since part of its power is not radiated. A more efficient pulse shape is the first order derivative of the Gaussian pulse, obtained making $n$=1 in expression 2.8 (it is necessary to normalize this signal to volt by multiplying it by $\sigma$). The pulse waveform and corresponding PSD are depicted in Fig. 2.12. The PSD has a fractional bandwidth equal to 1.66, $f_L$=90 MHz and $f_H$=990 MHz.

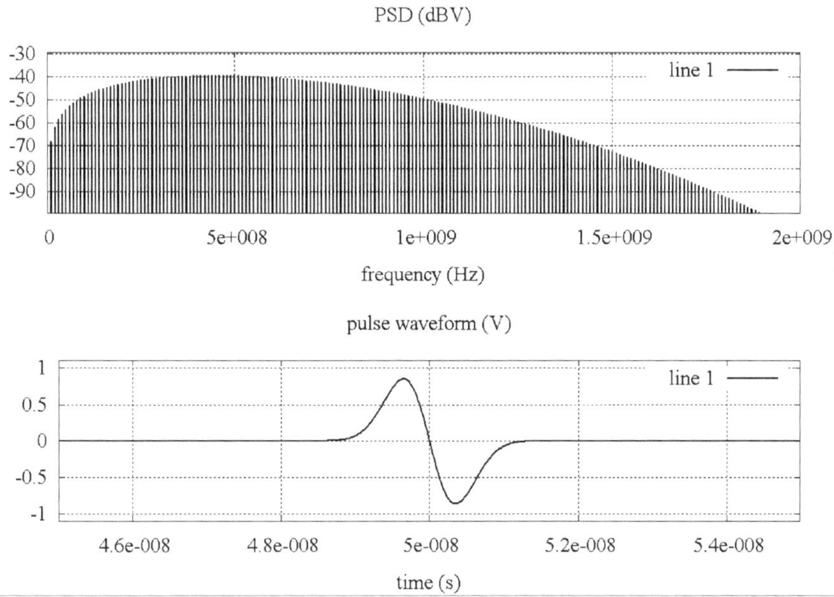


**Fig. 2.12** Power spectral density of the first derivative of a Gaussian pulse with $\sigma$=0.5 ns and PRF=10 MHz

The power of this signal is given by (considering that $T_{PR}$>>$\sigma$):

$$P = \frac{1}{T_{PR}} \cdot \int_{T_{PR}} v(t)^2 dt \approx A^2 \cdot \frac{\tau}{T_{PR}} \cdot \frac{\sqrt{2\times\pi}}{8} \qquad (2.10)$$

Another signal without DC component is the second order derivative of the Gaussian pulse, obtained by making $n$=2 in expression 2.8 (it is necessary to normalize this signal to Volts by multiplying it by $\sigma^2$). The pulse waveform and corresponding PSD are depicted in Fig. 2.13. The PSD has a fractional bandwidth equal to 1.31, $f_L$=290 MHz and $f_H$=1160 MHz. The PSD is shifted up in the frequency when compared to the previous signal.

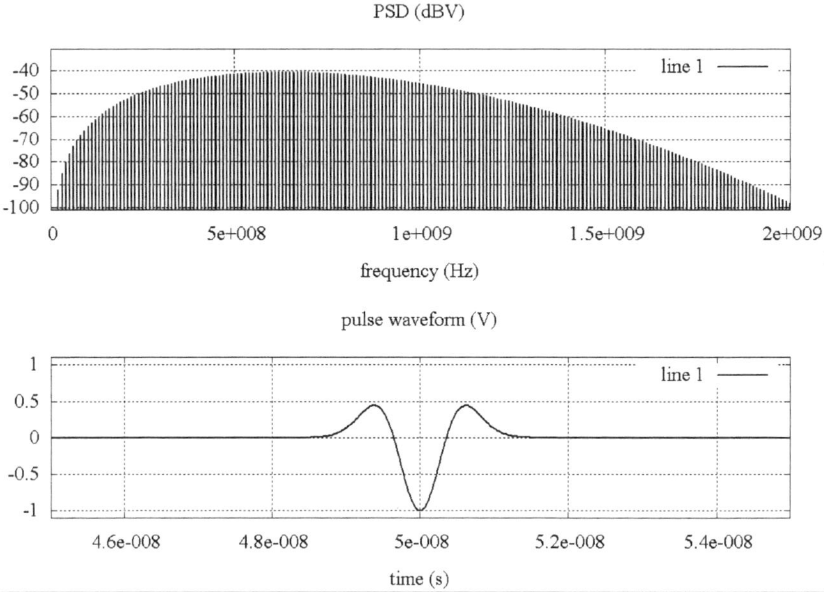


**Fig. 2.13** Power spectral density of the second derivative of a Gaussian pulse with $\sigma$=0.5 ns and PRF=10 MHz

The PSD of this signal has a smaller fractional bandwidth compared to the first derivative of the Gaussian pulse and it is higher in the frequency band. The power of this signal is given by (considering that $T_{PR}$>>$\sigma$):

$$P = \frac{1}{T_{PR}} \cdot \int_{T_{PR}} v(t)^2 dt \approx A^2 \cdot \frac{\tau}{T_{PR}} \cdot \frac{3 \times \sqrt{2 \times \pi}}{2} \tag{2.11}$$

## 2.3 Antennas for UWB Signals

The antenna converts the electrical signals to electromagnetic waves that propagate in space. This conversion is frequency dependent and therefore the antenna acts as a filter. Depending on the frequency value, the antenna attenuates the signal amplitude and changes the signal delay. This is not a problem for narrow band signals, since these signals can be considered approximately sinusoidal and therefore, their shape remains unchanged (sinusoidal). In fact, the amplitude and phase are altered by the antenna, but the signal remains a sine-wave with the same frequency. UWB signals on the other hand, can have their shape significantly changed by the frequency response of the antenna. Antennas can be divided into four basic types, considering their behavior as function of the frequency [39]:

1. *Electrically small antennas*: characterized by being much smaller than a wavelength in extent. They exhibit very low directivity,[4] low input resistance and high input reactance. Examples are the short dipole and small loop.
2. *Resonant antennas*: these antennas are designed to have real input impedance at the operating frequency and are characterized by having moderate gain at the operating frequency. They only maintain these characteristics in a narrow bandwidth around the resonant frequency. Examples are the half wave dipole and microstrip patch.
3. *Broadband antennas*: the gain and impedance remains nearly constant over a wide frequency range. These are characterized by having an active region where most of the power is radiated, the size of this region changes with frequency which results in a constant gain over the frequency range. The operating bandwidth of these antennas can span more than one decade. Examples are the spiral antenna and the log-periodic dipole array.
4. *Aperture antennas*: are characterized by having an opening through which electromagnetic waves flow. This aperture is usually several wave-lengths long in one or more dimensions. These antennas exhibit a high gain which increases with frequency. The operating bandwidth of these antennas can span more than one octave. An example is the horn antenna.

In order to understand how each of these antennas types is suited for the transmission of UWB signals, it is necessary to analyze how the antenna affects the UWB signals that is transmitting or receiving. Traditional antenna theory [41], analyzes the antenna behavior by solving the Maxwell's equations assuming harmonic fields (sinusoidal steady state solution). The resulting antenna parameters are defined for a single frequency. These parameters can be numerically calculated (using the method of the moments for example) or measured for different frequencies and then used to calculate the time domain waveforms through the inverse Fourier transform [42], [37]. Equivalent time-domain antenna parameters can also be defined [43], [44]. Alternatively the antenna transfer function can be calculated directly in the time-domain [45].

The value of the module of the electrical field produced by an antenna, in a point in the far field region,[5] in the direction defined by the $\theta$ and $\phi$ angles, is given by [39]:

---

[4] Directivity (in a given direction): the ratio of the radiation intensity in a given direction from the antenna to the radiation intensity averaged over all directions. (Definition from [40]).

[5] The far field (Fraunhofer) region is defined for $R >> \dfrac{2 \cdot L^2}{\lambda}$ [41] where $R$ is the radial distance from the antenna, $L$ is the largest dimension of the antenna and $\lambda$ is the wave-length. An equivalent condition can be defined for UWB signals: $R >> \dfrac{L^2}{c \cdot \lambda}$ [45], where $\tau$ is the UWB pulse duration.

$$E^{rad}(\omega,\theta,\phi,R) = -j \cdot Z_0 \cdot \beta \cdot \frac{I_{inp}(\omega)}{4 \cdot \pi \cdot R} \cdot he(\omega,\theta,\phi) \cdot e^{-j \cdot \beta \cdot R} \tag{2.12}$$

where $Z_0$ is the free space impedance, $\beta = \omega / c$ is the wave number, $I_{inp}$ is the current at the antenna input terminals, $R$ is the radial distance from the antenna to the point of measure and $he$ is the module of the effective antenna length vector (this value depends on the direction).

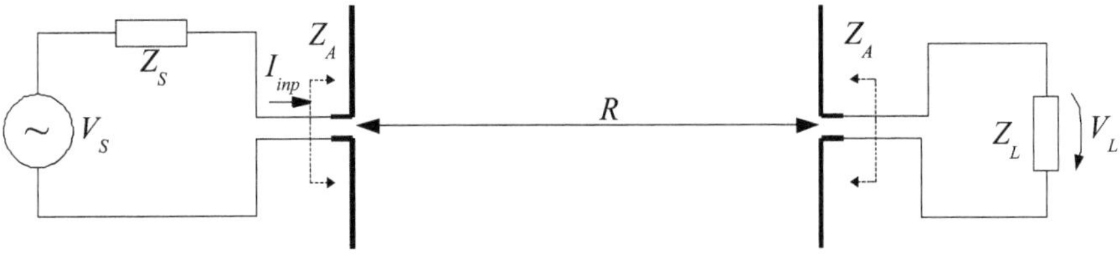


**Fig. 2.14** Diagram of a pair of two antennas aligned according to the maximum transmit direction and minimum polarization losses

The value of the current at the antenna terminals can be calculated from the values of the antenna input impedance ($Z_a$) and antenna input voltage ($V_s$) (from Fig. 2.14):

$$I_{inp}(\omega) = \frac{V_S(\omega)}{Z_a(\omega) + Z_S(\omega)} \tag{2.13}$$

where $Z_s$ is the impedance of the voltage source applied to the antenna. The gain of the antenna in a given direction is defined as "the ratio of the radiation intensity in a given direction to the radiation intensity that would be obtained if the power accepted by the antenna would be radiated isotropically" [40]. The power radiated by the antenna is dependent on the matching between the input source impedance and the antenna impedance, part of the available power from the source is radiated and the other part is reflected back to the source, the power radiated by an antenna in a given direction is:

$$P(\omega,\theta,\phi) = \frac{V_S(\omega)^2}{4 \cdot R_S} \cdot \left(1 - |\Gamma(\omega)|^2\right) \cdot G_t(\omega,\theta,\phi) \tag{2.14}$$

where $R_s$ is the value of the real part of the source impedance, $\Gamma$ is the power reflection coefficient and $G_t$ is the antenna gain. The EIRP value of an antenna is defined as [40]:

$$EIRP(\omega) = P_t(\omega) \cdot G_{t\,max}(\omega) \tag{2.15}$$

where $P_t$ is the power accepted by the antenna and $G_{t\,max}$ is the maximum antenna gain.

The open-circuit voltage in a receiving antenna is determined by the internal product between the incident electrical field vector and the effective length vector of the receiving antenna. Assuming that this vector is aligned to the incident electrical field (the incoming wave is coming from the maximum gain direction of the antenna) and assuming that there are no polarization losses (the incident electrical wave is aligned with the antenna effective length vector) the open-circuit voltage ($V_r$) is given by:

$$V_r(\omega) = E^i(\omega) \cdot he(\omega) \tag{2.16}$$

It is important to note the asymmetry between transmitting transfer function of an antenna and the receiving transfer function of the same antenna, respectively expressions (2.12) and (2.16), the radiated electrical field is proportional to the wave number ($\beta$). This means that the transmitting transfer function is proportional to the derivative of the receiving transfer function [38].

The previous expressions can be used to calculate the voltage across an impedance $Z_L$ connected to the receiving antenna (from Fig. 2.14):

$$V_L(\omega) = E^i(\omega) \cdot he(\omega) \cdot \frac{Z_L(\omega)}{Z_a(\omega) + Z_L(\omega)} \tag{2.17}$$

Assuming that $V_s(\omega)$, $Z_a(\omega)$ and $he(\omega)$ are known (for a particular direction and polarization), the time domain waveform of the voltage in the load, resulting from applying a single UWB pulse to the transmitting antenna, is given by the inverse Fourier transform:

$$v_L(t) = \int_{\omega=-\infty}^{+\infty} V_L(\omega) \cdot e^{-j\omega t} d\omega \tag{2.18}$$

The previous expressions suggest that it is possible to define the voltage transfer function of the antenna link (in a given direction and with a certain polarization) as [37]:

$$H(\omega) = \frac{V_L(\omega)}{V_S(\omega)} \tag{2.19}$$

### 2.3.1 Electrically Small Antennas

Considering as an example of an electrically small antenna, the infinitesimal dipole antenna (such as the ones depicted in Fig. 2.14) , which is characterized by its length ($l$) being much smaller than the wave-length. This antenna has an effective length equal to $\frac{l}{2}\cdot\sin(\theta)$ and its input impedance is constituted by a capacitance $\left(C_a = \frac{\pi}{2\cdot Z_0\cdot c}\cdot\frac{l}{\ln(l/2\cdot a)}\right)$ in series with a small resistance (the radiation resistance $R_r = 2\times\pi^2\cdot\left(\frac{l}{\lambda}\right)^2$ ). Considering that two infinitesimal dipoles are separated by a distance $R$, oriented so that there are no polarization losses and that the gain is maximum, the link voltage transfer function is given by:

$$H(\omega) = -j\cdot Z_0\cdot\frac{\omega}{c}\cdot\frac{e^{-j\cdot\beta\cdot R}}{4\cdot\pi\cdot R}\cdot\left(\frac{l}{2}\right)^2\cdot\frac{Z_L}{(Z_a+Z_S)\cdot(Z_a+Z_L)} \tag{2.20}$$

If the source impedance and the load impedances are purely resistive, with value equal to $R_S$ (much larger than $R_r$) and assuming that approximation $Z_a \approx 1/(j\cdot\omega\cdot C_a)$ is valid,[6] the previous transfer function has 3 zeros and 2 poles given by:

$$z_{1,2,3} = 0 \quad p_{1,2} = \frac{-1}{C_a\cdot R_S} \tag{2.21}$$

The transfer function has a triple zero at DC (as expected the antenna does not radiate at DC) followed by a double pole. The expression of the double pole is only valid if the value of the radiation resistance at the pole frequency is still much smaller than the value of $R_S$. This type of transfer function has the following implications into the shape of the received pulse: if the signal power is mainly located at frequencies well below the $p_{1,2}$ frequency, the received signal shape would be close to the third derivative of the transmit pulse. If the signal power is mainly located above the pole frequency it is not possible to conclude what is the signal shape because the previous approximation might not be valid and the antenna might not be considered a short dipole for those high frequencies.

In order to determine the exact shape the of the received signal it is necessary to calculate the values of $Z_a(\omega)$ and $he(\omega)$ in the frequency band occupied by the signal (using an antenna simulation program, such as MININEC, based on the

---

[6] Of course, this is only valid for signals with frequencies resulting in wave-lengths much smaller than l.

method of the moments [39], [46]) and then use expression (2.18) to determine the voltage across the load impedance (in reality, the IFFT algorithm is used to compute the inverse Fourier transform). The radiated electrical field at 10 m (obtained by applying the inverse Fourier transform to expression 2.12) and the load voltage are calculated for a pair of antennas, with 10 cm of length (this corresponds to a resonant frequency of approximately 1.5 GHz), separated by 10 m, when the input signal is a Gaussian pulse with $\sigma$=1.5 ns, the resulting waveforms are depicted in Fig. 2.15.

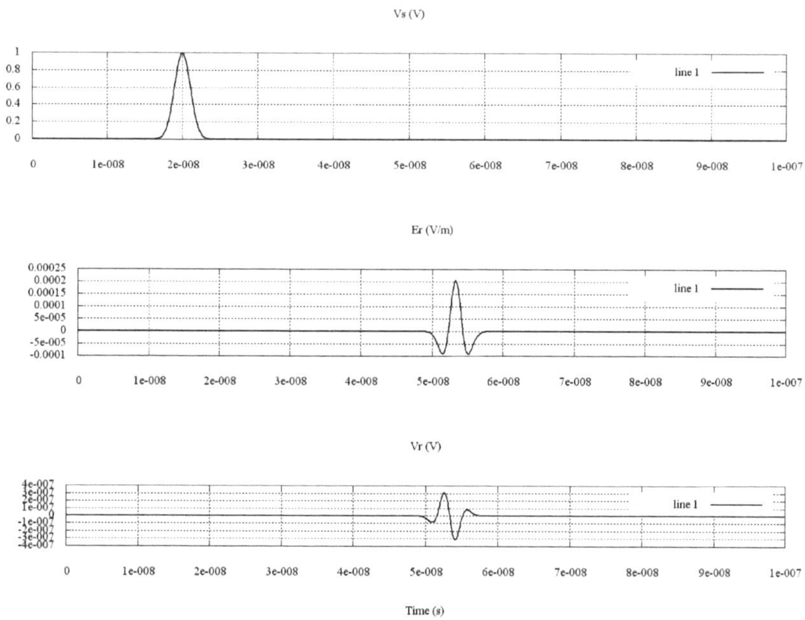


**Fig. 2.15** Waveform of an antenna link constituted by two dipole antennas, with 10 cm length, separated by 10 m, when a Gaussian impulse with $\sigma$=1.5 ns (top graph) is applied to the transmitting antenna through 50 Ω impedance. The middle graph is the module of the radiated electrical field at 10 m distance and the bottom graph is the voltage in the load impedance of the second antenna (50 Ω)

The voltage across the load impedance in the receiving antenna is, as expected, the third derivative of the voltage signal applied to the transmitting antenna. The transmitting antenna current is the first derivative of the input voltage because the antenna impedance is essentially capacitive. The radiated electrical field is the derivative of the antenna current and the voltage across the load impedance is the derivative of the receiving antenna open circuit voltage, because the antenna impedance is essentially capacitive. The electrical field produced by the transmitting antenna at 10 m distance, is the second derivative of the Gaussian pulse and it is delayed from the input pulse by $R/c \approx 33$ ns. The value of the received signal amplitude is extremely small (less than 1 μV). This is because the impedance of the antenna (essentially determined by $C_a$) is very large (much larger than the 50

Ω source impedance), which for an input signal with 1 V amplitude results in a low value for the current in the transmitting antenna (expression 2.13). The radiated field is therefore small (expression 2.12). The EIRP of the radiated signal is determined using expression 2.15, the PSD of the EIRP is shown in Fig. 2.16 together with the PSD of the input signal and the spectral mask defined in Table 2.1.

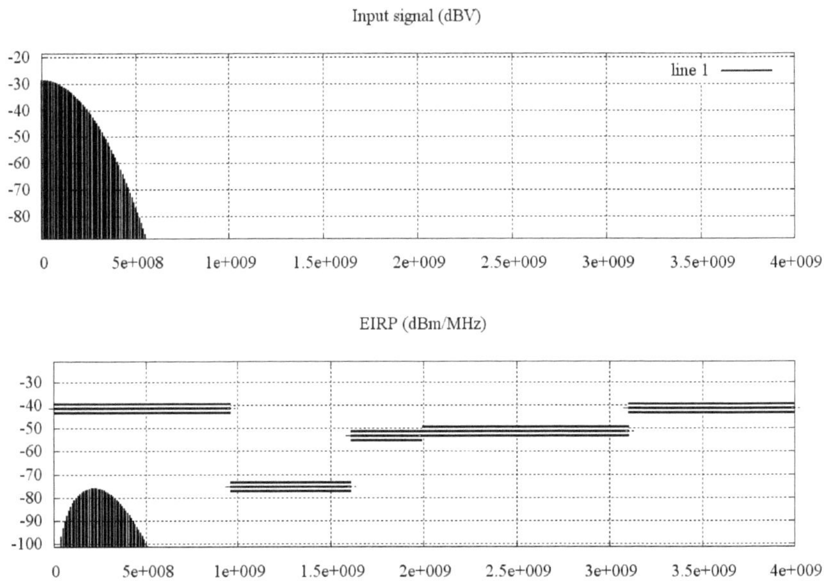


**Fig. 2.16** PSD of the input signal (in dBV) and EIRP of the radiated signal (dBm/MHz) by a 10 cm dipole

The EIRP maximum value is below the limit set in Table 2.1, in order to increase the PSD of the radiated signal to be closer to the limit, it would be necessary to increase the amplitude value of the input signal to nearly 50 V. In order to produce pulses with a duration around 1 ns and an amplitude of 50 V, it is necessary to use special devices such as avalanche transistors or step recovery diodes.

In the receiving antenna, the antenna impedance is also much larger than the load impedance and the voltage division between these two impedances results in a large attenuation of the open circuit voltage of the antenna. The power transfer from the input voltage source to the load impedance of the receiving antenna is also very small. If the values of the source and load impedances are chosen to be equal to the conjugate value of the antenna impedance (matched impedances), the power transfer between the input voltage source and the load of the receiving antenna is maximized. The problem with this method is that the value of the antenna impedance changes with the frequency of the signal and it is very difficult to create an impedance matching circuit that works over a large frequency band. This is best illustrated by the following example: using a source and load impedances constituted by a resistor (equal to $R_r$) in series with an inductance (equal to

$1/\left(\omega_{match}^2 \cdot C_a\right)$), results in matching these two impedances to the antenna impedance at a single frequency value ($\omega_{match}$). In this case the transfer function (2.20) is a 4th order band-pass filter with two pairs of complex conjugated poles located at $\omega_{match}$, resulting that the (wide-band) input pulse will be extremely distorted. This is shown in Fig. 2.17, in this case the source impedance and load impedance are matched to the antenna for a frequency of $\omega_{match}$=400 MHz.

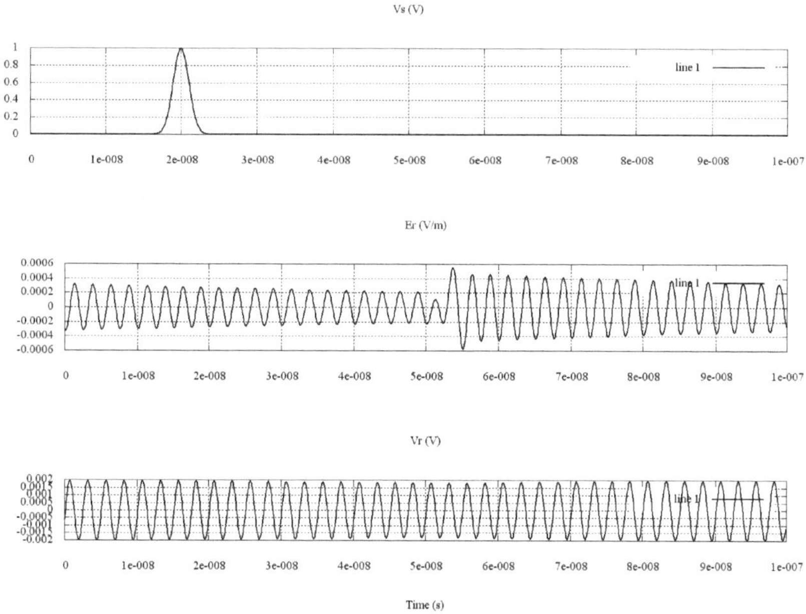


**Fig. 2.17** Waveform of an antenna link constituted by two dipole antennas, with 10 cm length, separated by 10 m, when a Gaussian impulse with $\sigma$=1.5 ns (top graph) is applied to the transmitting antenna through a conjugate impedance. The middle graph is the module of the radiated electrical field at 10 m distance and the bottom graph is the voltage in the load impedance of the second antenna (conjugate load)

The received signal amplitude is much larger now (approximately 2 mV), but the received signal is a damped sine-wave,[7] the damping factor is very small due to the large quality factor of the filter function (the quality factor is dependent on the series resistance value, which in this case is very small). Any information encoded in the input signal using the time domain (using its position for example) is lost. This suggests that the traditional approach of matching impedances, used in narrow band systems, might not be suited for UWB systems.

---

[7] The received signal starts before the input pulse because the IFFT assumes that the signals are periodic, with a period equal to the analysis time window. Thus, the oscillations occurring in the received signal before the arrival of the radiated signal are due to a pulse that occurred in the previous period.

## 2.3.2 Resonant Antennas

Resonant antennas are characterized by having physical dimensions that are directly related to the wavelength of the signal. At this frequency (or in a small bandwidth around this frequency) the gain is maximum and the antenna impedance is real. When the wavelength of the electromagnetic waves is different from this optimal wavelength the gain (in a given direction) is reduced and the antenna impedance is no longer real. Therefore, these types of antennas are bandpass systems and as such, will distort UWB signals. To understand this distortion, a dipole antenna (such as the ones depicted in Fig. 2.14) with a length equal to 25 cm, corresponding to a half-wavelength for a frequency approximately equal to 600 MHz, was used to perform the same analysis described before. Using a Gaussian signal with a $\sigma$=1 ns as the antenna input voltage the resulting waveforms are depicted in Fig. 2.18.

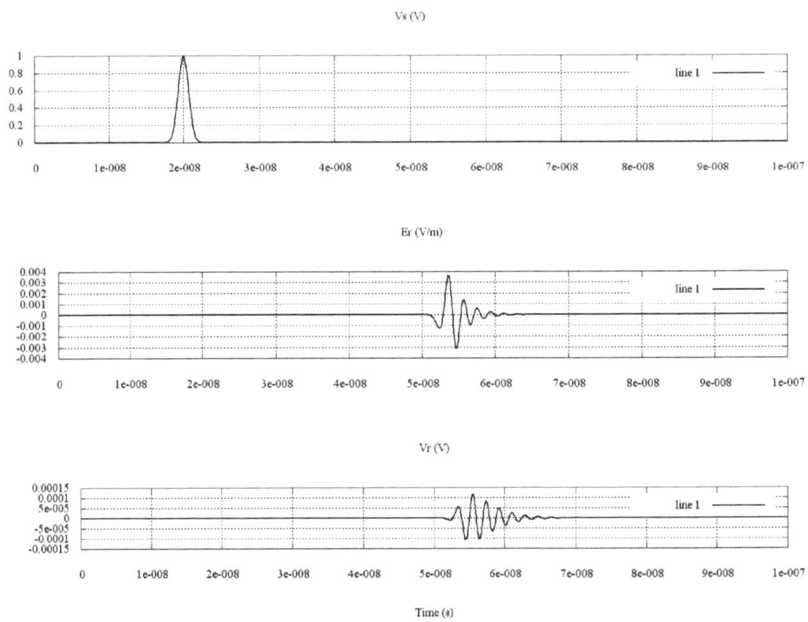


**Fig. 2.18** Waveform of an antenna link constituted by two dipole antennas, with 25 cm length, separated by 10 m, when a Gaussian impulse with $\sigma$=1 ns (top graph) is applied to the transmitting antenna through a 50 Ω resistance. The middle graph is the module of the radiated electrical field at 10 m distance and the bottom graph is the voltage in the load impedance of the second antenna (50 Ω load)

The received signal is very similar to a damped sine-wave, the input signal is concentrated in approximately 1 ns and the received signal is spread around approximately 15 ns. This can be a problem for many UWB systems, since the uncertainty of the signal position has increased 10 times.

The PSD of the radiated signal is shown in Fig. 2.19; the amplitude of the input signal is set to 3 V, in order to obtain the maximum allowed value for the EIRP. This amplitude value is still achievable using low cost CMOS technologies.

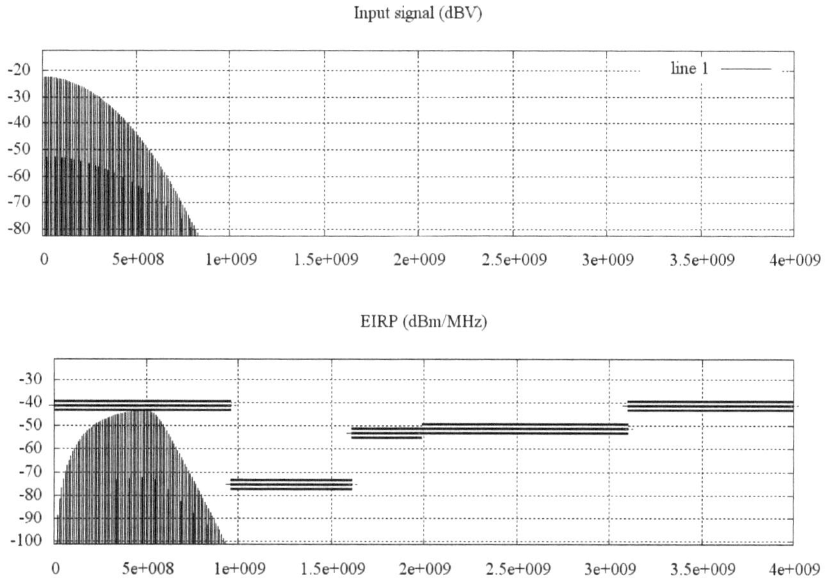


**Fig. 2.19** PSD of the input signal (in dBV) and EIRP of the radiated signal (dBm/MHz) by a 25 cm dipole. The amplitude of the input signal is 3 V

A technique commonly used to extend the bandwidth of resonant antennas is distributed resistive loading [21], [38]. This technique consists in using a material with a high resistivity to build the dipole antenna. The distributed resistance causes a rapid decay in the current in the antenna structure and hence a wide bandwidth is achieved at the price of a reduced efficiency. This is akin to increasing the bandwidth of an RLC circuit by increasing the resistance value. To illustrate this technique, the previous dipole was analyzed considering that the dipole had total distributed resistance of 62 Ω. The resulting waveforms are depicted in Fig. 2.20.

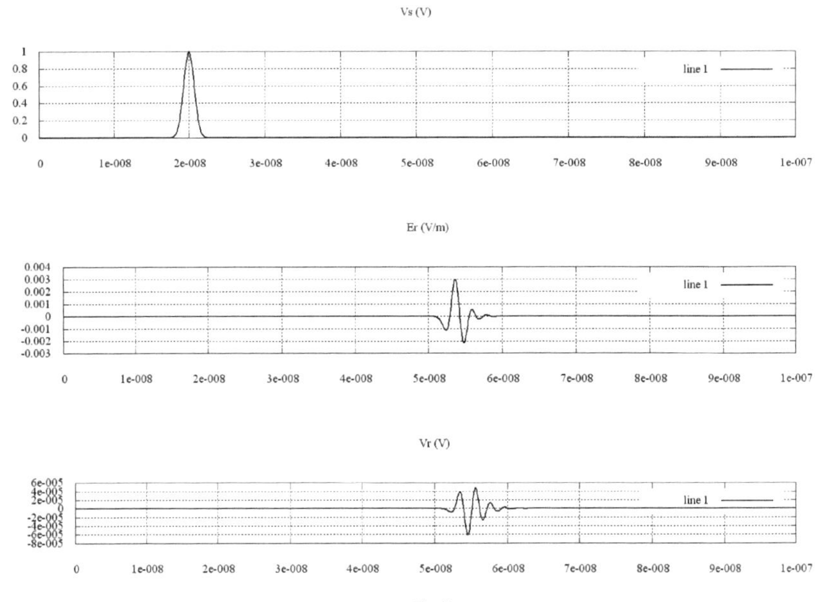


**Fig. 2.20** Waveform of an antenna link constituted by two resistively loaded dipole antennas (with a total distributed load equal to 62 Ω), with 25 cm length, separated by 10 m, when a Gaussian impulse with σ=1 ns (top graph) is applied to the transmitting antenna through a 50 Ω impedance. The middle graph is the module of the radiated electrical field at 10 m distance and the bottom graph is the voltage in the load impedance of the second antenna (50 Ω load)

The received signal energy is now concentrated in a smaller time interval (around 5 ns) and its waveform is closer to the third derivative of a Gaussian pulse. But the maximum amplitude of the received signal was reduced from 100 μV to 60 μV. The PSD of the EIRP of the radiated signal is depicted in Fig. 2.21, this was obtained after the input signal amplitude was increased to 4 V in order to achieve the maximum allowed EIRP level. The resistively loaded dipole can be an useful antenna for the reception of UWB signals, but it is very inefficient as a radiator of UWB signals due to its high impedance value across the frequency, that result in a low radiated power unless a high voltage signal is used at the input. The dipole resistance will also generate extra thermal noise that will reduce the sensitivity of the receiver circuit.

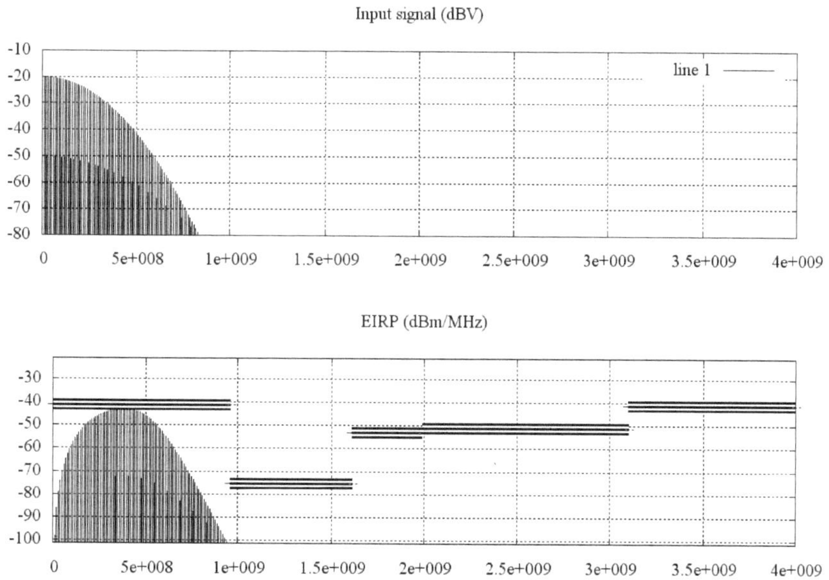


**Fig. 2.21** PSD of the input signal (in dBV) and EIRP of the radiated signal (dBm/MHz). The amplitude of the input signal is 4 V

## *2.3.3 Broadband Antennas*

The characteristics of an antenna, such as impedance and gain, are dependent on the physical dimensions of the antenna and the frequency of the signal. If the antenna physical dimensions are scaled by a given factor *k*, the performance of the antenna will remain unchanged if the signal frequency is scaled by 1/*k*. Therefore broadband antennas require physical structures that are characterized by having an emphasis on angles rather than on lengths [39]. This leads to a self scaling behavior: most of the radiation occurs in a part of the antenna (the so called active region) with a width equal to half a wave-length or a circumference equal to a wavelength. As the frequency decreases, the active region size increases and becomes closer to the size of the antenna. The resulting antenna exhibits (approximately) constant gain and input impedance across the frequency band, but the effective area of the antenna changes with frequency. Examples of structures exhibiting such characteristics are: bi-conical antennas (constituted by two infinite inverted cones), bow-tie antennas (which is a planar version of the bi-conical antenna) and spiral antennas.

In both the bi-conical and the bow-tie antennas the signal is feed to the tips of inverted cones (or triangles). The characteristics of these antennas are determined by the angle of the cone (or triangle) $\alpha$ in Fig. 2.22.

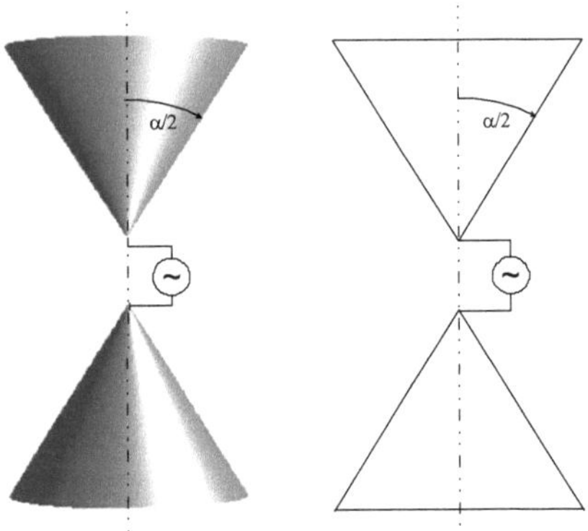


**Fig. 2.22** Bi-conical antenna (left) and bow-tie antenna (right)

The impedance of an infinite bi-conical antenna is constant across frequency and is given (for free space) by [46]:

$$Z_a = \frac{Z_0}{\pi} \cdot \ln\left[\cot\left(\frac{\alpha}{4}\right)\right] \tag{2.22}$$

which is purely resistive, an input impedance of 50 Ω is obtained with an angle $\alpha$ around 134°. In practice, the bi-conical and the bow-tie antenna must have finite dimensions, which causes a reflected traveling wave at the end the of the cone (or triangle). This causes standing waves whose frequencies are related to the dimension of the antenna; these standing waves create a complex part in the antenna impedance and make the antenna impedance frequency dependent. The bi-conical antenna becomes a very impractical structure to be built for low frequency signals because of the large solid structure (or shell) needed for the cones. The bi-conical antenna can be approximated by several wires, resulting in a broadband performance close to the solid antenna. This structure, depicted in Fig. 2.23, was simulated using MININEC and the values obtained for the antenna impedance and effective height were used to calculate time domain waveforms and PSD of the UWB signals, these waveforms are shown next in Fig. 2.24.

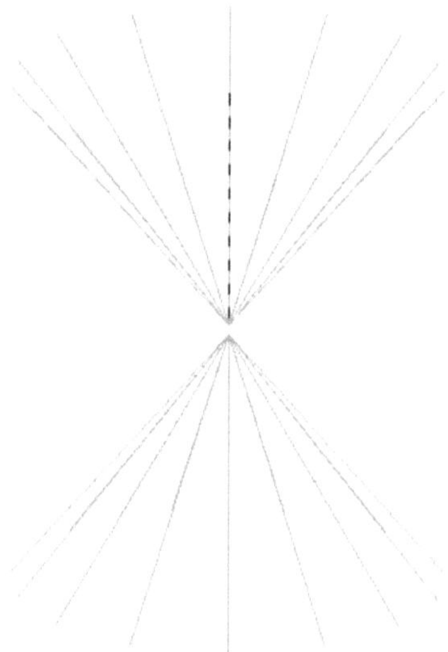

**Fig. 2.23** Bi-conical antenna built using wires with 20 cm length and an angle $\alpha$ equal to 90°

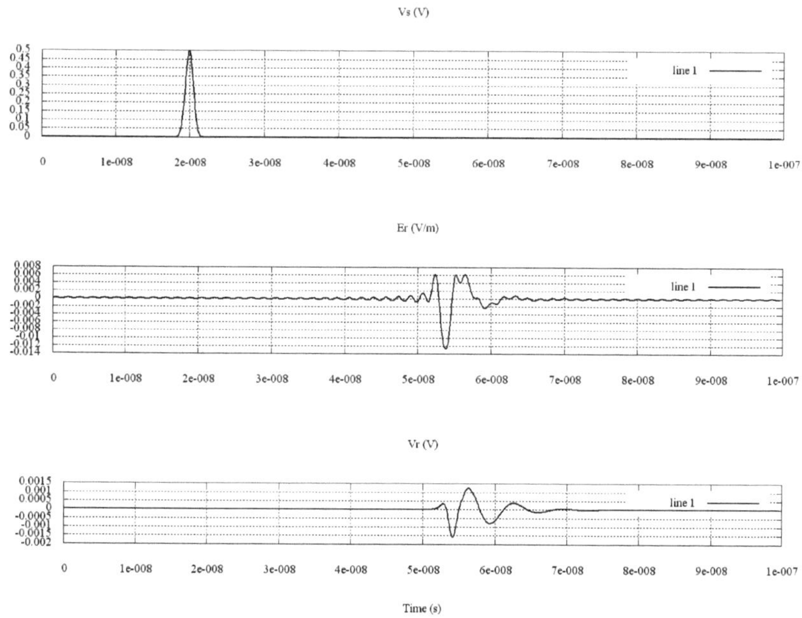


**Fig. 2.24** Waveform of an antenna link constituted by two resistively loaded bi-conical antennas, built using several 20 cm length wires, separated by 10 m, when a Gaussian impulse with $\sigma$=0.75 ns (top graph) is applied to the transmitting antenna through a 50 Ω impedance. The middle graph is the module of the radiated electrical field at 10 m distance and the bottom graph is the voltage in the load impedance of the second antenna (50 Ω load)

The finite bi-conical antenna characteristics are not constant with frequency; they vary with the signal frequency, oscillating between maximum values and minimum values. The antenna impedance is not real for all frequencies; the impedance exhibits several resonant frequencies, a behavior similar to a circuit composed by several RLC circuits. This variation causes the signal power to be higher at certain frequencies and lower at others frequencies. This causes the resulting time waveform to exhibit ringing with different frequency values. The frequency values of this ringing depend on the antenna and on the input signal PSD. This type of antenna has a larger efficiency for UWB signal than the resonant antennas; the radiated electrical field is approximately the third derivative of the input signal and has an amplitude much larger than the one obtained with the dipole antenna. The received signal pulse has a low frequency ringing, but its amplitude is large. This ringing can be attenuated by using a capacitor as the load impedance ($Z_L$) to attenuate the low frequency ringing, the waveform for this case are shown in Fig. 2.25.

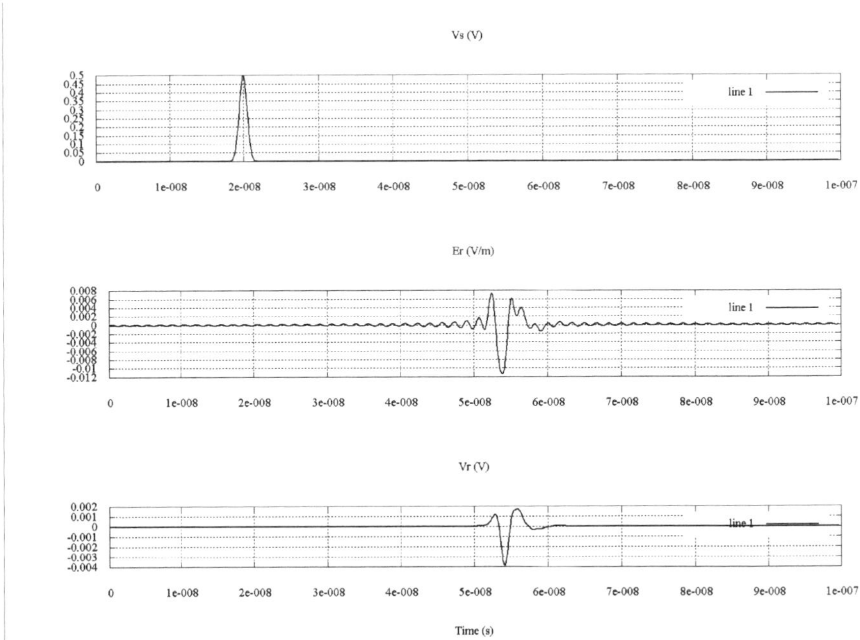


**Fig. 2.25** Waveform of an antenna link constituted by two resistively loaded bi-conical antennas, built using several 20 cm length wires, separated by 10 m, when a Gaussian impulse with $\sigma$=0.75 ns (top graph) is applied to the transmitting antenna through a 50 $\Omega$ impedance. The middle graph is the module of the radiated electrical field at 10 m distance and the bottom graph is the voltage in the load impedance of the second antenna (1 pF load)

The PSD of the radiated signal by a finite bi-conical antenna, for an input signal with 0.5 V amplitude is shown in Fig. 2.26. The PSD of the radiated signal is more uniform across the frequency band than the PSD of a signal radiated by a dipole antenna.

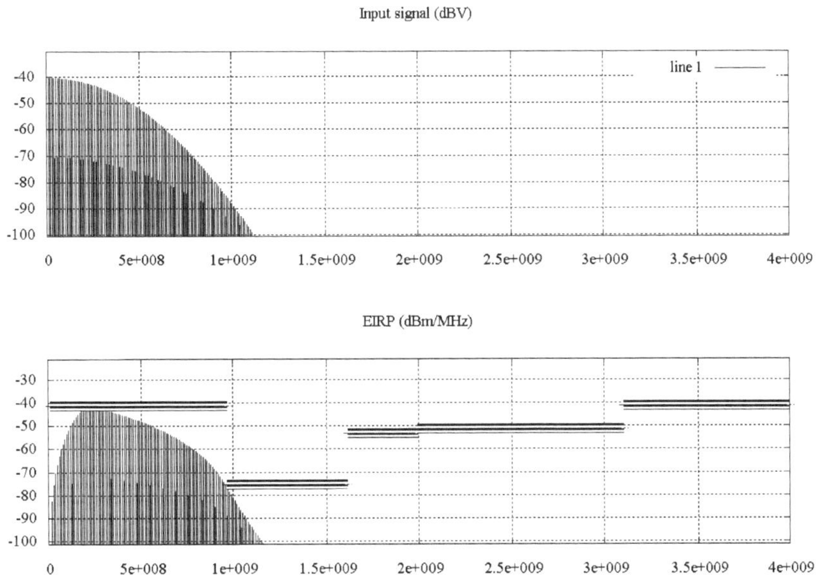


**Fig. 2.26** PSD of the input signal (in dBV) and EIRP of the radiated signal (dBm/MHz). The amplitude of the input signal is 0.5 V

Another type of broadband antenna is the spiral antenna [47], constituted by two spirals that unwind with a given flare rate. The flare rate can be exponential or constant, in which case the resulting antenna is known as an Archimedes antenna. Since spiral antennas are 2-D structures they can be easily constructed using PCB techniques. An example of an Archimedes antenna is depicted in Fig. 2.27.

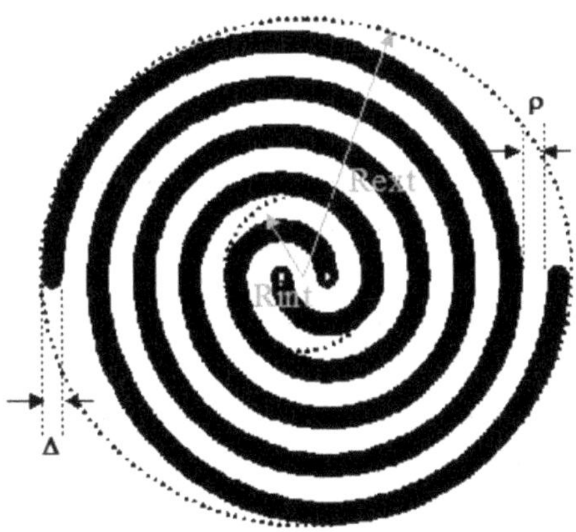


**Fig. 2.27** Archimedes spiral antenna

These antennas are designed to be self-complementary. The spiral arms are feed using a balanced source and thus the current in the beginning of the spiral arms starts in opposite directions. As the current travels in the spiral arms its phase is delayed. When the current travels exactly half a wave-length in the spiral arm the phase is delayed 180°. Radiation occurs in the antenna region where the current in both spiral arms is in phase, i.e., when the length of one spiral arm is equal to one or more wave-lengths and the length of the other spiral arm is equal to one or more wave-length plus half a wave-length. This region is confined to an area with a perimeter equal to one wave-length; beyond this active region, the currents are small because most of the power is radiated in the active region. The perimeter of the radiation region reduces as the signal frequency increases. This is illustrated, in Fig. 2.28, where the magnitude of the current in the spiral arms is shown for two signals with frequencies 800 MHz and 2400 MHz. In the case of the higher frequency signal, the current magnitude is very small in outer region of the antenna because the signal power is almost completely radiated in the center region of the antenna.

The antenna is designed to have an internal circumference perimeter (defined by $R_{int}$) equal to the minimum wave-length desired and an external circumference perimeter (defined by $R_{ext}$) equal to the maximum wave-length desired. The antenna can have a large bandwidth (over 1 decade) with reasonable physical dimensions

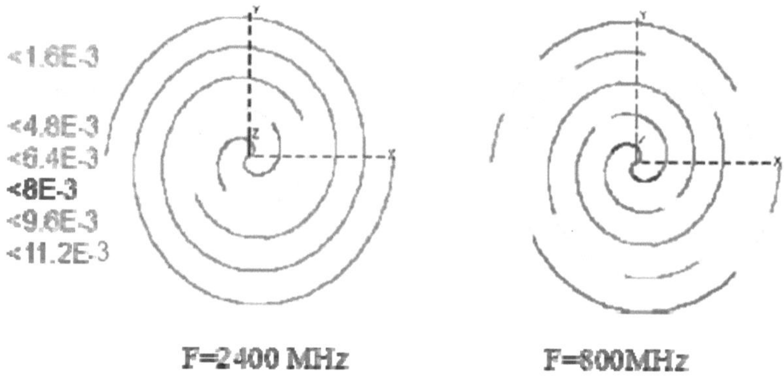


**Fig. 2.28** Magnitude of the current in the antenna spiral arms for two different signal frequencies

However, when an UWB signal is applied to the antenna feed point, which is located in the center of the antenna, the high frequency components of the signal are radiated first and the low frequency components of the signal are radiated latter. This is because, as the signal travels through the antenna arms, the high frequency signals are radiated almost completely in the center region of the antenna and the low frequency components of the signal are radiated when the pulse arrives at the end of the antenna arms. This phenomenon can distort the shape of the UWB signal into a chirp form [21], [48], resulting in an increase of the pulse duration (dispersion). This makes the spiral antenna not indicated for UWB signals.

## 2.3.4 Aperture Antennas

Aperture antennas have an opening or aperture in their structure, through which electromagnetic waves flow. Due to this physical aperture the effective area of the antenna is constant (but not equal to the physical aperture). Aperture antennas are highly directional and have a large gain that varies with the signal frequency. The TEM horn antenna is constituted by two conducting plates whose spacing and width are increased until a rectangular opening is formed, the angle between the plates (flare angle) can be constant or it can change along the width of the antenna. This antenna is feed at the extremity where the spacing between the plates is smallest. The antenna structure is depicted in Fig. 2.29

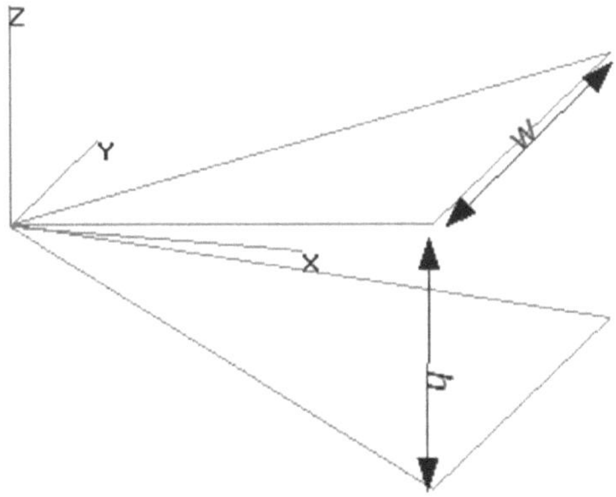


**Fig. 2.29** Simplified Horn antenna layout. The antenna aperture is given by $w \times h$

The impedance along the $x$ dimension can be viewed as the impedance of a waveguide, with two plates with a given width ($w(x)$) and a given spacing ($h(x)$). The objective is to slowly change the impedance value, from the value of the impedance of the transmission line used to apply the signal to the antenna, to the value of the impedance of free space. The antenna guides the TEM mode present in the transmission line into free space radiation. The signal is radiated when the spacing between the plates ($h(x)$) is equal to half a wave-length. The minimum signal frequency radiated by the antenna is determined by the spacing between of the plates at the end of the antenna, therefore the antenna can be considered as a high pass filter. The flare angle should be small in order to provide a smooth transition between the line impedance and the free space impedance and thus obtain a low VSWR [38]. If the minimum desired signal frequency is small then the spacing between the plates at the end of the antenna is large and the antenna length must also be large to provide a smooth transition from the transmission line to the large opening at the end. The antenna's dimensions can be reduced if continuous resistive loading is introduced into the conducting plates [38], however, this solution decreases the antenna efficiency. Another option is to exponentially flare the angle between the two plates. Then, it is possible to obtain an antenna with a smaller length and still have low VSWR across the frequency band. This type of antenna is known as the exponential horn antenna [49]. The exponential horn antenna is still fairly large for low frequency applications, the antenna presented in [49], designed to cover a frequency band from 0.2 GHz to 2 GHz is 76 cm long, 94 cm wide and 69 cm high. The horn antenna presented in [50] is designed to cover the frequency band from 1 GHz to 11 GHz, has a depth of 13 mm and an aperture of 25 mm × 20 mm, these small dimensions are obtained by terminating the launching plate with resistors in order to reduce the reflections that occur at the end of the launching plate.

Horn antennas have been used to transmit and receive UWB signals. In [38] the author analyzes the time domain waveforms of UWB signals in rectangular TEM horn antennas with and without resistive loading. These waveforms show that this type of antenna does not distort or disperse significantly the UWB signals. In [50] a pyramidal exponential horn antenna with a bandwidth between 1 GHz and 11 GHz is analyzed, the time domain waveforms in this antenna show that the UWB signals are not significantly distorted. When a second order derivative of the Gaussian pulse is applied the received signal is similar to the third derivative of the Gaussian signal.

### 2.3.5 Antenna Summary

An UWB signal that is radiated will always have its shape distorted since the radiation acts as differentiation. Depending on the antenna type the pulse shape can be further distorted and dispersed in time.

Electrically small antennas do not introduce a significant dispersion into the pulse shape, but this type of antenna introduces a very strong attenuation into the UWB and it is only suitable for very small ranges requiring an UWB signal generator with a large voltage amplitude.

From the previous analysis it is clear that resonant antennas are not indicated for UWB signals since they tend to distort and introduce significant time dispersion into UWB signals. This dispersion can be reduced by resistively loading the antenna, but this reduces the efficiency.

Certain broadband antennas, such as the conical antenna, can be well adapted to UWB signals, since they have a reasonable gain and do not distort significantly the signals.

Aperture antennas such as the horn antenna can be designed to have a large bandwidth and therefore accommodate UWB signals without significant distortion or dispersion. This type of antenna has the advantage of being very directional.

All the antennas have in common: that if the minimum operating frequency is lower the maximum physical dimensions increase. This can be improved by introducing loading resistors at certain key places in the antenna structure to reduce reflections that cause standing waves.

Since these types of antennas (non-loop antennas) have a capacitive impedance at low frequencies, the shape of UWB signals is better preserved if the antenna is loaded with a capacitive impedance, since this type of loading will lower the dispersion of the received UWB signal.

## References


1. Constable, A.R.; The birth pains of radio; International Conference on 100 Years of Radio, 1995, 5–7 Sep. 1995 Page(s):14–19
2. Bryant, J.H.; The first century of microwaves-1886 to 1986; IEEE Transactions on Microwave Theory and Techniques, Volume 36, Issue 5, May 1988 Page(s):830–858
3. Thrower, K.R.; History of tuning; International Conference on 100 Years of Radio, 1995, 5–7 Sep. 1995 Page(s):107–113
4. Armstrong, E.H.; Some Recent Developments in the Audion Receiver; Proceedings of the IRE, Volume 3, Issue 3, Sep. 1915 Page(s):215–238
5. Armstrong, E.H.; A Study of Heterodyne Amplification by the Electron Relay; Proceedings of the IRE, Volume 5, Issue 2, April 1917 Page(s):145–159
6. Armstrong, E.H.; the Super-Heterodyne-Its Origin, Development, and Some Recent Improvements; Armstrong, E.H.; Proceedings of the IRE, Volume 12, Issue 5, Oct. 1924 Page(s):539–552
7. Ross, G.F.; The evolution of UWB technology; Radio and Wireless Conference, 2003. RAWCON '03. Proceedings 10–13 Aug. 2003 Page(s):1–4
8. Bennett, C.L.; Ross, G.F.; Time-domain electromagnetics and its applications; Proceedings of the IEEE, Volume 66, Issue 3, Mar. 1978 Page(s):299–318
9. Ross, G.F.; The Generation of Pulse-Modulated Signals at C Band and Beyond (Correspondence); IEEE Transactions on Microwave Theory and Techniques, Volume 19, Issue 1, Jan. 1971 Page(s):96–99
10. XYZs of oscilloscopes; Tektronix; Application Note
11. Time Domain Reflectometry Theory; Hewlett Packard; Application Note 1304-2
12. Using TDR to Help Solve Signal Integrity Issues; Tektronix; Application Note
13. Zhifang, Du; Willett, P.K.; Mashikian, M.S.; Performance limits of PD location based on time-domain reflectometry; IEEE Transactions on Dielectrics and Electrical Insulation, Volume 4, Issue 2, April 1997 Page(s):182–188
14. Zhifang Du; Mashikian, M.S.; Palmieri, F.; Partial discharge propagation model and location estimate; Electrical Insulation, 1994., Conference Record of the 1994 IEEE International Symposium on 5–8 June 1994 Page(s):99–102
15. Harney, W.; Nemarich, C.; Electromagnetic Level Indicating (EMLI) System Using Time Domain Reflectometry; OCEANS, Volume 15, Aug. 1983 Page(s):233–236
16. Nemarich, C.P.; Time domain reflectometry liquid level sensors; IEEE Instrumentation & Measurement Magazine, Volume 4, Issue 4, Dec. 2001 Page(s):40–44
17. Ross, G.F.; The Accurate Measurement of Range by the Use of Microwave Delay Line Techniques; MTT-S International Microwave Symposium Digest, Volume 75, Issue 1, May 1975 Page(s):341–342
18. Toevs, J.W.; Smith, D.R.; Byrne, D.P.; Ross G.F.; UWB radar for ground vehicle self protect; IEEE MTT-S International Microwave Symposium Digest, 1992., volume 3, 1–5 June 1992 Page(s):1491–1494
19. James, D. Taylor and Thomas E. McEwan; The Micropower Impulse Radar; Chapter 6 of "Ultra-wideband Radar Technology", Edited by James D Taylor; CRC Press; ISBN: 0849342678

20. Kotzebue, K.L.; A circuit model of the step-recovery diode; Proceedings of the IEEE, Volume 53, Issue 12, Dec. 1965 Page(s):2119–2120
21. Daniels, D.J.; Gunton, D.J.; Scott, H.F.; Introduction to subsurface radar; IEE Proceedings Radar and Signal Processing, F Volume 135, Issue 4, Aug. 1988 Page(s):278–320
22. Daniels, D.J.; Ultrawideband radar detection of buried objects; Geoscience and Remote Sensing, 1997. IGARSS '97. 'Remote Sensing - A Scientific Vision for Sustainable Development'., 1997 IEEE International, Volume 3, 3–8 Aug. 1997 Page(s):1278–1281
23. Daniels, D.J.; Brooks, D.; Dittmer, J.; Mitchell, O.; Hunt, N.; Wide swathe multi-channel GPR systems for mine detection; RADAR 2002 15-17 Oct. 2002 Page(s):210–216
24. Earp, S.L.; Hughes, E.S.; Elkins, T.J.; Vickers, R.; Ultra-wideband ground-penetrating radar for the detection of buried metallic mines; IEEE Aerospace and Electronic Systems Magazine, Volume 11, Issue 9, Sept. 1996 Page(s):30–39
25. Fontana, R.J.; Recent system applications of short-pulse ultra-wideband (UWB) technology; IEEE Transactions on Microwave Theory and Techniques, Volume 52, Issue 9, Part 1, Sept. 2004 Page(s):2087–2104
26. Gordon E. Moore; Cramming More Components onto Integrated Circuits; Electronics, Volume 38, Number 8, April 19, 1965.
27. Behzad Razavi; RF Microelectronics; Prentice Hall, ISBN 0-13-887571-5
28. Win, M.Z.; Scholtz, R.A.; Impulse radio: how it works; IEEE Communications Letters, Volume 2, Issue 2, Feb. 1998 Page(s):36–38
29. Mitchell, T.; Broad is the way [ultra-wideband technology]; IEE Review, Volume 47, Issue 1, Jan. 2001 Page(s):35–39
30. Liuqing, Y.; Giannakis, G.B.; Ultra-wideband communications: an idea whose time has come; IEEE Signal Processing Magazine, Volume 21, Issue 6, Nov. 2004 Page(s):26–54
31. Roy, S.; Foerster, J.R.; Somayazulu, V.S.; Leeper, D.G.; Ultrawideband radio design: the promise of high-speed, short-range wireless connectivity; Proceedings of the IEEE, Volume 92, Issue 2, Feb. 2004 Page(s):295–311
32. Revision of Part 15 of the Comission's rules regarding ultra-wideband transmission systems; Federal Communications Commission, First Report and Order FCC 02-48,ET docket 98–153, April 2002.
33. A. Bruce Carlson; Communication Systems; McGraw Hill, ISBN 0-07-100560-9
34. Taha, A.; Chugg, K.M.; On the power spectral density of wireless multiple-access UWB impulse radio under realistic propagation conditions; 2003 IEEE Vehicular Technology Conference, 2003. VTC 2003-Fall. 58th Volume 2, 6–9 Oct. 2003 Page(s):1298–1302 Vol.2
35. Shanmugan, K.S.; Estimating the power spectral density of ultra wideband signals; IEEE International Conference on Personal Wireless Communications, 2002 15–17 Dec. 2002 Page(s):124–128
36. Power Spectra Estimation; National Semiconductor, Application Note 255 November 1980.
37. Zhi, N.C.; Xuan, H.W.; Hui, F.L.; Ning, Y.; Chia, M.Y.W.; Considerations for source pulses and antennas in UWB radio systems; IEEE Transactions on Antennas and Propagation, Volume 52, Issue 7, July 2004 Page(s):1739–1748
38. Kanda, M.; Time domain sensors for radiated impulsive measurements; IEEE Transactions on Antennas and Propagation, Volume 31, Issue 3, May 1983 Page(s):438–444
39. Warren L. Stutzman, Gary A. Thiele; Antenna Theory and Design; Jonh Willey and Sons Inc.; ISBN 0-471-02590-9
40. IEEE standard definitions of terms for antennas; IEEE Std 145-1983 22 June 1983
41. Balanis, C.A.; Antenna theory: a review; Proceedings of the IEEE, Volume 80, Issue 1, Jan. 1992 Page(s):7–23
42. Franceschetti, G.; Papas, C.; Pulsed antennas; IEEE Transactions on Antennas and Propagation, Volume 22, Issue 5, Sep. 1974 Page(s):651–661
43. Lamensdorf, D.; Susman, L.; Baseband-pulse-antenna techniques; IEEE Antennas and Propagation Magazine, Volume 36, Issue 1, Feb. 1994 Page(s):20–30

44. Baum, C.E.; E.G. Farr; C.A. Frost; Transient gain of antennas related to the traditional continuous-wave (CW) definition of gain;Baum, Ultra-Widebańd Short-Pulse Electromagnetics 4, 1998 14-19 June 1998 Page(s):109–118
45. Heyman, E.; Kastner, R.; Antenna characterization in the time domain; Shlivinski, A.; IEEE Transactions on Antennas and Propagation, Volume 45, Issue 7, July 1997 Page(s):1140–1149
46. Balanis, C.A; Antenna Theory; Jonh Willey and Sons Inc.; ISBN 0-471-59268-4
47. Dyson, J.; The equiangular spiral antenna; IEEE Transactions on Antennas and Propagation, Volume 7, Issue 2, April 1959 Page(s):181–187
48. Hansen, R.C.; Libelo, L.F.; Wideband dispersion in baseband systems; IEEE Transactions on Aerospace and Electronic Systems, Volume 31, Issue 3, July 1995 Page(s):881–890
49. Kerr, J.; Short axial length broad-band horns; IEEE Transactions on Antennas and Propagation, Volume 21, Issue 5, Sep. 1973 Page(s):710–714
50. Xu Li; Hagness, S.C.; Choi, M.K.; van der Weide, D.W.; Numerical and experimental investigation of an ultrawideband ridged pyramidal horn antenna with curved launching plane for pulse radiation; Antennas and Wireless Propagation Letters, Volume 2, Issue 1, 2003 Page(s):259–262

# Chapter 3
# UWB RADAR Receiver Architecture

**Abstract** this chapter describes the operation of a radar system. The differences and advantages of using UWB signals in the radar system, over traditional narrow band signals, are discussed. The radar equation, usually defined for narrow band signals, is redefined for UWB signals. This new radar equation is used to analyze the echo signals from targets with basic shapes, resulting in an estimative of the echo signal amplitude as a function of the target distance. The architecture of an UWB radar based in the concept of sub-sampling is defined and the constituting blocks are described. The remaining of the chapter is concentrated in analyzing the receiver channel of the radar system. This analysis is centered around the noise power level in the receive path. From this analysis a new receiver circuit for base-band UWB signals is proposed. This new circuit is an averaging switched-integrator circuit. This circuit is described in detail and a design procedure is shown.

## 3.1 UWB RADAR Systems

The operation principle of a pulsed RADAR[1] system is very simple [1] an electromagnetic signal (pulse) is radiated, this signal travels at the speed of light (*c*) (assuming that the signal propagates in space). When the signal encounters an object (target) part of its energy is reflected back (re-radiated), therefore creating an echo, the echo signal travels back at the speed of light to the point of the radiation of the original signal. The radar can use a single antenna for transmission and reception of the pulse or two separate antennas. The instant of arrival of the echo signal can be determined and the time interval (*T*) between the transmission of the signal and arrival of the echo can be measured. Fig. 3.1 shows a simple diagram illustrating the radar operation principle.

[1] **RADAR** stands for **RA**dio **D**etection **A**nd **R**anging.



N. Paulino et al., *Low Power UWB CMOS Radar Sensors*, 49–114.

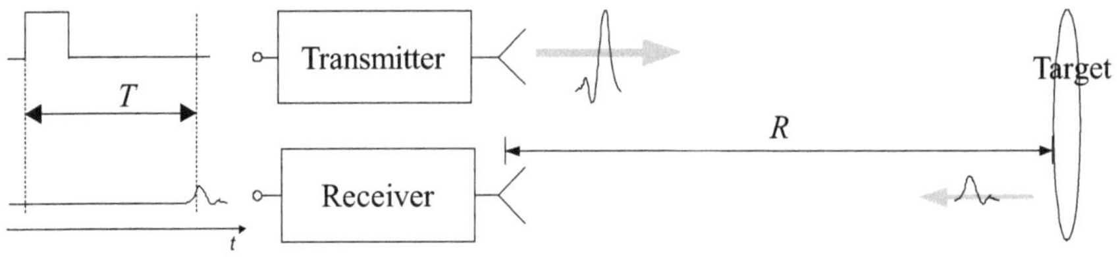


**Fig. 3.1** Simple diagram showing the principle of radar operation. The signals on the right side of the figure are the transmit clock and the received signal. The time interval T is measured from the rising edge of the transmit clock, which determines the pulse radiation, to the arrival instant of the echo signal

The distance (or range) (*R*) at which the target is located is calculated using the following simple formula

$$R = \frac{c \cdot T}{2} \tag{3.1}$$

If two targets are located at the same range, but located in different directions, their echo signals will arrive at the same time. This can be useful, for example in an intrusion alarm, where the objective is to determine if anything has penetrated a restricted area, but if the objective is to identify and locate different targets in different directions, it is not desirable. Depending on the objective, the radar system might use an antenna with low directivity (first case) that radiates over 360°, or an antenna with a large directivity that concentrates most of the radiated power in a narrow beam-width (second case), this type antenna attenuates the echo signals that are generated by targets located outside the radiation beam of the antenna, allowing targets located in different directions to be differentiated.

A single pulse produces a single echo from the target (or several echoes from several targets), the information contained in the echo is used to determine the position of the target (or targets) when the original pulse was transmitted. In order to obtain the position of the target at a different time instant, it is necessary to transmit a new pulse; a RADAR system usually transmits pulses periodically in order to track moving targets. The time interval between two transmitted pulses should be large enough to allow for all the echo signals created by the first pulse to be received. If a second pulse is radiated before all the echoes from the first pulse arrive, the next received echo can be from a target located at a close range or an echo created by the first pulse from a target located at a distant range and it is not possible to distinguish which case occurred. The time interval between two pulses must increase as the potential targets are located further away. The maximum Pulse Repetition Frequency (*PRF*) of a radar system is dependent on the maximum distance at which a target can be detected. If, for example, the maximum range at which a target might be detected is 300 Km, the maximum value for the *PRF* would be limited to 500 Hz, but if the maximum range is 30 m, the maximum

*PRF* value can be as high as 5 MHz. The maximum unambiguous range of the RADAR ($R_{max}$) is given by

$$R_{\max} = \frac{c \cdot T_{PRF}}{2} \tag{3.2}$$

where $T_{PRF}$=1/*PRF*. Obviously, there will always exist echoes generated from targets located further away than the maximum range value given by expression (3.2). These echoes are indistinguishable[2] from the echoes generated by targets inside the maximum range. Usually the PRF value is selected so that echoes from outside the maximum RADAR range have an amplitude smaller than the minimum detectable amplitude by the RADAR receiver.

Since the target can reflect the transmitted signals with different amplitudes (and phase shifts) at different frequencies, depending on the target shape and size, the echo signal can have a different shape and longer duration than the radiated pulse (pulse dispersion). Considering that two targets are illuminated by a transmitted pulse, the signal received by the receiver would be the combination of the echoes produced by the two targets. Any two targets separated by a radial distance from the RADAR inferior to $c.\tau$ (where $\tau$ is the pulse width) produce a combined echo signal that can be interpreted as the one created by a single target. Using signal processing techniques to separate two targets in theses conditions can be a very difficult or even impossible task. As a consequence the value of the pulse duration ($\tau$) determines the resolution that the RADAR can distinguish two targets in a given direction. This is illustrated in Figs. 3.2 and 3.3.

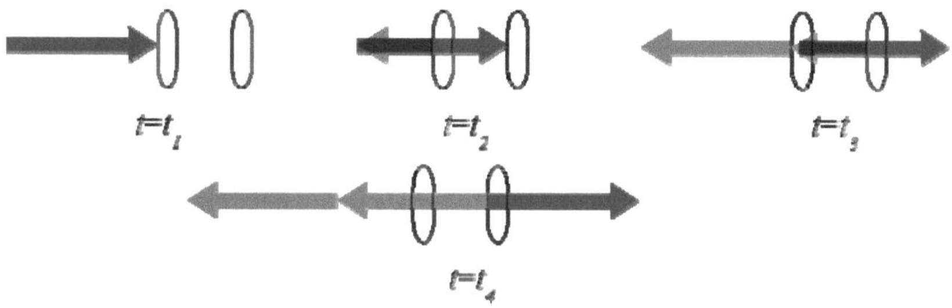


**Fig. 3.2** Simple diagram showing two targets separated by a distance smaller than $c.\tau$, illuminated by a pulse at different time instants. The transmitted pulse is represented by the dark arrow and the echo signals by the light arrows. The two echo signals are very difficult to separate from each other and will be interpreted as a single echo by the receiver

[2] If the PRF value varies it is possible to distinguish echoes from targets outside the RADAR maximum range.

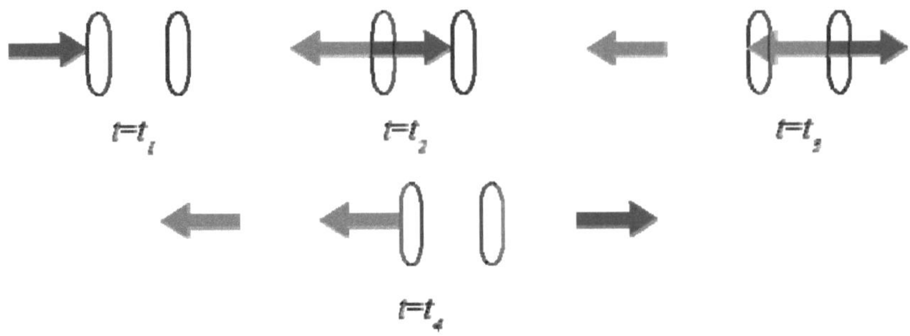


**Fig. 3.3** Simple diagram showing two targets separated by a distance larger than c.τ, illuminated by a pulse at different time instants. The pulse is represented by the dark arrow and the echo signals by the light arrows. The two echo signals can be clearly separated from each other (assuming that the targets do not introduce significant dispersion into the pulses and increase their duration)

It should be noted that reducing the value of $\tau$ will reduce the energy per pulse (considering constant pulse amplitude), and therefore reduce the energy of the each echo signal, making it more difficult to detect the target echoes. Increasing the resolution of a radar system usually result in a reduction of the maximum radar range. Traditional radar systems use a sine-wave (carrier) pulse, the pulse contains enough sine-wave cycles to be considered a narrow band signal (it has a fractional bandwidth inferior to 10%). This type of signal is similar to the one depicted in Fig. 2.7 of the previous chapter. In order to fall under this narrow band condition when using short duration pulses, it is necessary to use a high carrier frequency. As an example, to identify targets separated by a distance smaller than 30 cm it is necessary to use a pulse with a duration inferior to 1 ns, this results that a carrier frequency larger than 20 GHz must be used in order to achieve a fractional bandwidth inferior to 10% [2], [3]. Using a carrier signal with such a high frequency, requires that the circuits that process this signal must be constructed using expensive RF technologies, such as GaAs [4]. If carrier frequencies higher than 40 GHz are used then the radar system must be built using discrete components such as GUNN and Schottky diodes mounted on expensive substrates [5], [6].

In order to reduce the maximum signal frequency, while maintaining the same target separation capability in the radar system, it is necessary to decrease the carrier frequency while maintaining the same pulse duration; this corresponds to increasing the signal fractional bandwidth. The absolute minimum carrier frequency possible is zero, i.e. use short duration pulses without a carrier (UWB signals) such as the ones described in the previous chapter. Using UWB signals has also the advantage of producing an echo signal with more information about the target. Due to the large bandwidth occupied by this type of signal, several resonant frequencies of the target can be activated. The fact that the pulse width is very small can also be used to determine the shape of targets larger than $c.\tau$, since multiple echoes can be obtained from such a target. It should be noted that extracting this extra information about the target from the echo signal can require complex signal processing techniques and depending on the target and on the environment conditions might not be possible.

By employing UWB signals it is possible to use CMOS technology to implement the radar transceiver. CMOS technology has the advantage of having a lower cost (especially if a high production volume is required) and ease of integration into larger digital systems, when compared with other more expensive RF technologies. Current low cost CMOS technologies (0.35–0.18 μm) have lower maximum operating frequencies than GaAs or Silicon-Germanium technologies, but as smaller geometry CMOS technologies become available at lower prices, such as the 90 nm technology, the performance gap between low-cost CMOS technology and other more expensive technologies will be reduced [7], [8].

The disadvantage of using base-band short pulses is that the received echo signals from these pulses do not have the same shape as the pulse that was initially radiated; this is due to the distortion caused by the antennas and by the multiple reflections created by the target and environment. Another aspect is that the antennas must have a large bandwidth to accommodate the bandwidth of the UWB signals. If the antenna bandwidth is smaller than the signal bandwidth, the antenna will severely distort the signal's shape (as explained in the previous chapter). The distortion of the pulse shape makes it difficult to use matched filter circuits, since these filters have an impulse response equal to the time-reversed pulse shape. A matched filter is a circuit that maximizes the probability of detecting the presence of a pulse within noise at a given time instant; these types of circuits are extensively used in radar systems [9]. In a narrow band radar the received pulse shape is defined by the envelope of the carrier, which is not significantly distorted, since it is defined by the amplitude of the sine-wave (the carrier signal).

The amount of noise generated by a circuit is proportional to its bandwidth, the larger the bandwidth the larger the noise generated by the circuit (assuming constant impedance). Therefore circuits with a large bandwidth (needed to accommodate UWB signals) have an inherent high noise level and thus a low sensitivity. Since the amount of power allowed to be radiated is limited for UWB signals (as described in the previous chapter), the energy level of the return echoes is low. These two factors result in a shorter range for a UWB radar system when compared to a traditional radar system. UWB radar systems are more suited for short range high resolution applications, where the lower cost of a CMOS UWB system (compared to a narrow band radar system) is an advantage [10].

The use of low cost CMOS technologies, such as 0.35–0.18 μm CMOS technologies, limits the minimum pulse width produced by the UWB transceiver to values between 0.5 ns and 1.5 ns, limiting the application of CMOS technology to UWB signals with frequencies below 2 GHz. This limit is enough to cover the several imaging systems defined by the FCC rules, such as Ground Penetrating Radar Systems, Wall Imaging Systems and Through-wall Imaging Systems. All these systems can operate using signals with a PSD up to 960 MHz, corresponding to UWB signals with a time duration of approximately 1 ns.

## 3.2 UWB Radar Equation

The relation between the transmitted signal amplitude, target characteristics, target range and received signal amplitude is given by the radar equation. The radar equation relates the amount of transmitted power by the radar with the amount of power received by the radar reflected by the target at a given range (echo signal). A simplified diagram of the radar system and target, containing the relevant signals, is shown in Fig. 3.4.

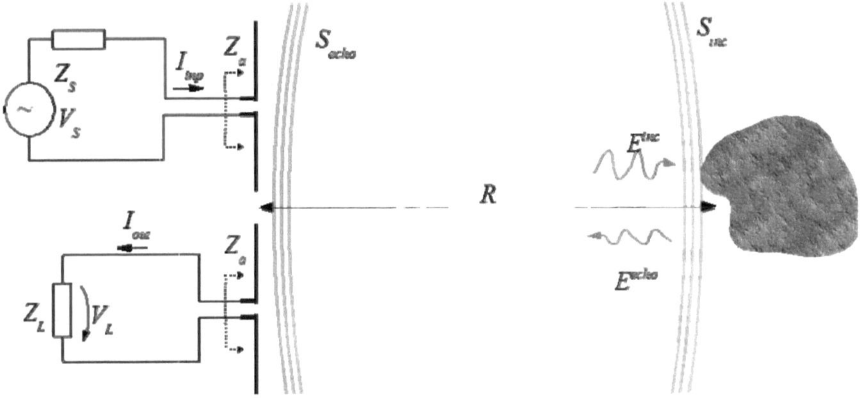


**Fig. 3.4** Simplified diagram of a monostatic radar system (in this case represented by a transmit and receive antenna with the same characteristics) and the target

It is important to know the expected amplitude of the received signal echo signal in order to design the receiver circuit. The minimum expected amplitude will determine the required value for the maximum receiver gain and the required value for the receiver input referred noise; the maximum expected amplitude is important to verify the linearity of the receiver circuit.

In the case of a narrow band radar system, the radar equation is derived considering that the radar is transmitting a continuous sine-wave signal with a fixed frequency. This is of course an approximation, since in reality the radar is transmitting pulses composed of several sine-wave cycles. The previous approximation is only valid for narrow band signals, where each pulse contains enough sine-wave cycles to be considered a continuous sine-wave. In the case of UWB signals this approximation can not be applied and it is necessary to reinterpret the radar equation deduction.

The narrow band radar equation is obtained [1] considering that the transmitter is radiating a continuous sine-wave signal with a power equal to $P_t$, if the transmitted electromagnetic wave propagates isotropically, then the power flux density at a target located at a range $R$ from the radar is given by

$$S_{inc}(R) = \frac{P_t}{4 \cdot \pi R^2} \tag{3.3}$$

The power flux density is simply the transmitted power divided by the area of a sphere with a radius equal to $R$. If the transmitter uses an antenna with gain $G_t$, aligned with the direction of the target, the previous expression can be rewritten as

$$S_{inc}(R) = \frac{P_t \cdot G_t}{4 \cdot \pi R^2} \tag{3.4}$$

Part of this incident power is re-radiated back towards the radar by the target. The amount of reflected power, by the target, back to the radar, depends on the target characteristics (size, geometry and material) and is determined by the radar cross-section of the target ($\sigma$). A target with a larger radar cross-section reflects more power back to the radar than a target with a smaller radar cross-section. Although the radar cross-section of the target is defined as an area, it does not have, necessarily, a direct relation with the area of the side of the target that is illuminated by the incident wave. The radar cross-section changes with the signal frequency and with the side of the target that is illuminated. The target can also change the polarization of the reflected wave, affecting in that way the amount of power received by the radar antenna. The power density of the reflected wave, by the target, at the radar location is given by

$$S_{echo}(R) = \frac{P_t \cdot G_t}{4 \cdot \pi R^2} \cdot \frac{\sigma}{4 \cdot \pi R^2} \tag{3.5}$$

Finally, the power received by the radar from the reflected wave, is dependent on the effective aperture ($A_e$) value of the reception antenna in the radar and it is given by

$$P_r(R) = \frac{P_t \cdot G_t}{4 \cdot \pi R^2} \cdot \frac{\sigma \cdot A_e}{4 \cdot \pi R^2} \tag{3.6}$$

The previous expression is a simplified form of the radar equation. A more complete form would include the radar receiver sensitivity (the minimum amount of power necessary for the radar receiver to detect a target echo) and various losses in the signal path. All the parameters in the expression are determined for the operating frequency of the radar, assuming a steady state analysis for the carrier signal (sine wave). In the case of an UWB signal, which occupies a very large bandwidth, these parameters can change significantly within the signal bandwidth, making the previous expression useless to estimate the received echo amplitude.

In order to obtain a valid expression for the echo amplitude in the case of UWB signals it is necessary to analyze the transient response of the system formed by the radar and the target, instead of the steady state solution for a sine-wave input [11]. As shown in the previous chapter, the transient response of a pair of linked antennas can be obtained from the frequency response of the antennas using the inverse Fourier transform. The objective is to determine the relation between the amplitude of the transmitted pulse ($V_S$) and of the received echo signal ($V_L$) as function of the signal frequency (assuming sine-wave signals). Once this transfer function is obtained, the time domain signals (transient response) can be calculated using the inverse Fourier transform and the amplitude of the received echo signal determined. This methodology allows reusing most of the previous analysis that led to the narrow band radar equation.

The transmitting antenna produces an electromagnetic wave that travels towards the target at the speed of light ($c$). The incident electrical field on the target, located in the direction of maximum gain of the antenna and at a distance $R$ from the antenna, has an amplitude given by

$$E^{inc}(\omega,R) = -j \cdot Z_0 \cdot \beta \cdot \frac{I_{inp}(\omega)}{4 \cdot \pi \cdot R} \cdot he(\omega) \cdot e^{-j \cdot \beta \cdot R} \tag{3.7}$$

The target scatters the incident electromagnetic radiation, producing new electromagnetic waves that radiate away from the target. The target radar cross-section is defined in [12] as the area ($\sigma$) that multiplied by the power flux density of the incident plane wave, results in the power necessary for an isotropic radiator to produce the same radiation intensity as the target, in a given direction (usually the direction of the reception antenna) and with the polarization of the reception antenna. The value of the radar cross-section depends on [13] the shape of the target, the constituting material of the target, the angle of illumination on the target, the frequency of the incident radiation and the polarization of the transmitting and receiving antennas. The radar cross-section value is defined as [1]

$$\sigma = \lim_{R \to \infty} 4 \cdot \pi \cdot R^2 \cdot \left| \frac{E^{echo}(R)}{E^{inc}(R)} \right|^2 \tag{3.8}$$

where $E^{echo}(R)$ is the electrical field scattered by the target at the radar location and $E^{inc}(R)$ is the electrical field produced by the radar at the target location.

By rearranging expression 3.8, it is possible to obtain the magnitude of the scattered electrical field (with the same polarization as the receive antenna) at the target location

$$E^{echo}(\omega,0) = \sqrt{\frac{\sigma(\omega)}{4 \cdot \pi}} \cdot \left| E^{inc}(R) \right| \tag{3.9}$$

This expression can be used to obtain the voltage amplitude of the sine-wave signal, with frequency $\omega$, received by the radar, using

$$V_L(\omega,R) = \left| E^{echo}(\omega,R) \cdot he(\omega) \cdot \frac{Z_L(\omega)}{Z_L(\omega) + Z_a(\omega)} \right| \tag{3.10}$$

However, since the objective is to determine the time domain representation of the received voltage for an UWB signal, it would be necessary to calculate the inverse Fourier transform of expression 3.10. But this is not possible because this expression only provides the module of the Fourier transform of the received signal. The phase of the signal is not available because the radar cross-section of the target is defined as the ratio between the power of the reflected wave and the power of the incident wave (expression 3.8).

In order to obtain the received voltage in the time domain it is necessary to know the relation between the phase of the incident sine-wave signal and the corresponding echo sine-wave signal reflected by the target, as a function of the signal frequency ($\omega$). It is also necessary to know the value of the module of the target radar cross-section as a function of the signal frequency. The radar cross-section of a target can also be defined as [13]

$$\sigma = \left| \sum_{k=1}^{N} A_k \cdot e^{j \cdot \phi_k} \right|^2 \tag{3.11}$$

where each $A_k$ and $\phi_k$ are associated with a particular scattering source in the target. Assuming that the target is linear and produces a reflected signal without harmonics,[3] it is possible to combine the previous expression with expression 3.9 resulting in

$$E^{echo}(\omega, R) = \frac{1}{\sqrt{4 \cdot \pi}} \cdot \sum_{K=1}^{N} A_K(\omega) \cdot e^{j \cdot \phi_k(\omega)} \cdot E^{inc}(\omega, R) \cdot \frac{e^{-j \cdot \beta \cdot R}}{R} \tag{3.12}$$

The complex exponential factor represents the phase shift that occurs in the sine-wave signal as it propagates in space with speed $c$. Combining this expression with expression 3.7 results in the following expression for the received voltage amplitude (as function of frequency)

$$V_L(\omega, R) =$$
$$-j \cdot Z_0 \cdot \beta \cdot \frac{I_{inp}(\omega)}{4 \cdot \pi \cdot R^2} \cdot he(\omega)^2 \cdot e^{-2 \cdot j \cdot \beta \cdot R} \cdot \frac{1}{\sqrt{4 \cdot \pi}} \cdot \sum_{K=1}^{N} A_K(\omega) \cdot e^{j \cdot \phi_k(\omega)} \cdot \frac{Z_L(\omega)}{Z_L(\omega) + Z_a(\omega)} \tag{3.13}$$

The received voltage in the time domain can be obtained by applying the inverse Fourier transform to the previous expression. The resulting expression can predict the voltage across the load impedance of the reception antenna as a func-

[3] An example of a non-linear target is described in [15] a small dipole antenna loaded by a resonant loop and Schottky diode. This target was designed and optimized in order to be attached to insects and facilitate their tracking by producing an echo signal with the double of the frequency of the incident signal.

tion of time. In the case when the transmitted signal is a pulse with a short duration, the received voltage will be equal to zero most of the time, differing from zero volt only around the time when an echo signal arrives at the radar.

In order to calculate the received echo voltage as a function of time ($v_L(t)$), it is necessary to know the transmitted pulse shape ($v_s(t)$), the antennas characteristics ($he(\omega)$ and $Za(\omega)$ of both the reception and transmitting antennas) and the target characteristic (the values of $A_k(\omega)$ and $\phi_k(\omega)$). In the following analysis it will be considered that the radar uses Gaussian pulses and a bi-conical antenna, such as the one analyzed in the previous chapter, for both transmission and reception. The amplitude of the Gaussian pulses is set so that the EIRP value is below the level specified in Table 2.1. The frequency content of the radiated wave is shown in Fig. 2.26 of the previous chapter. The only remaining unknowns are the target characteristics (the values of $A_k(\omega)$ and $\phi_k(\omega)$).

The radar cross-section of a given target can be determined by solving the Maxwell equations with the correct boundary conditions. However, this calculation can be very complicated resulting that it is only possible to determine analytically the radar cross-section for simple target shapes [1], for more complex target shapes, it is usually determined experimentally [16]. Therefore, the target shapes used in this analysis will be simple shapes constituted by a high conductance material, (metal). According to Snell's law, an infinite plane of metal reflects completely an incident perpendicular electromagnetic wave and the resulting reflected wave has an electrical field with an amplitude value equal to the negative amplitude of the incident wave electrical field. In the case of a finite plane of metal, only part of the incident power would be reflected, but the electrical field amplitude would still be inverted, resulting in the following target characteristic, for a perpendicular incident wave

$$|A(\omega)| = \sqrt{\sigma(\omega)} \qquad \phi = 180 \tag{3.14}$$

where $\sigma(\omega)$ is the finite plane radar cross-section value. A finite plane creates only one reflection, a target with a more complex shape can create more reflections and therefore it is necessary to use different $A_k(\omega)$ and $\phi_k(\omega)$ to determine its reflected wave.

Depending on the relation between the size of the target and the wavelength of the incident electromagnetic wave, there are different phenomenons that can cause the reflection and hence a different type of analysis should be used to estimate the value of the radar cross-section of the target depending on the frequency of the incident wave.

If the maximum target dimension ($l$) is smaller than the wavelength ($d_{\max} < \lambda$) it is considered that reflection of the incident wave is due to Rayleigh scattering[4]

[4] This is the phenomenon that causes the sky to appear to be blue; the molecules in the atmosphere are much smaller than the wavelength of the visible and therefore the light is scattered

and the target is referred as being in the Rayleigh region of the incident wave. The radar cross-section is inversely proportional to $\lambda^4$ and proportional to the squared target volume ($V^2$). In the case of a sphere with a circumference value equal to $2 \cdot \pi \cdot r$, the radar cross-section value is given by [13]

$$\sigma_{sphere1} = \frac{4}{\pi} \cdot \left( \frac{2 \cdot \pi}{\lambda} \right)^4 \cdot \left( \frac{4 \cdot \pi \cdot r^3}{3} \right)^2 \cdot \left( \frac{9}{4} \right)^2 \tag{3.15}$$

If the wavelength of the incident wave is much smaller than the maximum target dimension $d_{\max} > 10 \times \lambda$, it is considered that the reflection of the incident wave is due to optical reflection and target is referred as being in the optical region of the incident wave. In the case of a sphere, the radar cross-section value in the optical region is given by [13].

$$\sigma_{sphere2} = \pi \cdot r^2 \tag{3.16}$$

If the relation between the wavelength of the incident wave and the maximum dimension of the target is between the two limits indicated before, resonance phenomenon are responsible for the reflection and the target is referred as being in the resonance region of the incident wave. In this case it is not easy to obtain a simple expression for the radar cross-section value of a given target. The radar cross-section value can be larger than the value in the optical region and can vary a lot with the incident wavelength variation.

Figure 2.26. in Chapter 2, shows that the incident wave power, for a Gaussian pulse with 0.75 ns duration, is mainly located inside the 230 MHz–830 MHz frequency interval. Therefore a sphere with a circumference perimeter smaller than 0.34 m (corresponding to a radius smaller than 0.05 m) is considered to be in the Rayleigh region of the incident wave and its radar cross-section value can be calculated using expression 3.15. If the sphere has circumference larger than 13 m (corresponding to a radius larger than 2 m) the sphere is in the optical region of the incident wave and expression 3.16 can be used to calculate the radar cross-section value of the sphere.

A rectangular metal plate can be considered to be in the optical region of the incident wave, if the perimeter is larger than 13 m. The radar cross-section of a rectangular metal plate with dimension $a \times b$, assuming that the plate is perpendicular to the incident wave, can be calculated using [13], [14]

$$\sigma_{rec}(f) = \left( \frac{2 \cdot \pi \cdot a \cdot b}{c \cdot f} \right)^2 \cdot \frac{1}{\pi} \tag{3.17}$$

---

according to Rayleigh law ($\lambda^4$). The light with a smaller wavelength (blue light) is more scattered than the light with a larger wavelength (red light) and thus the sky appears to be blue.

Using the previous expressions it is possible to estimate the target characteristics for each target shape resulting in

$$\begin{aligned} |A_{sphere}(\omega)| &= \sqrt{\sigma_{sphere}(\omega)} \\ \phi_{sphere} &= 180 \end{aligned} \tag{3.18}$$

$$\begin{aligned} |A_{rec}(\omega)| &= \sqrt{\sigma_{rec}(\omega)} \\ \phi_{rec} &= 180 \end{aligned} \tag{3.19}$$

The received echo voltage for each target as a function of time can now be determined by calculating the inverse Fourier transform of expression 3.13 for each target shape (using the corresponding expression (3.18 or 3.19)). The received echo voltage is equal to zero until the time necessary for the pulse to travel between the radar and the target elapses twice and the actual echo signal appears. The shape of the echo signal depends on the transmitted pulse shape, on the target and on the antennas. The amplitude of the echo depends on the target range ($R$). In Fig. 3.5, the received echo voltages for two different targets located at a range of 10 m from the radar are shown. The two targets are a rectangular metal plate with 6 by 6 m and a metal sphere with a diameter equal to 5 cm. The transmitted pulse is a Gaussian pulse with 0.7 ns of duration and an amplitude equal to 0.4 V. The receive antenna is loaded by a 1 pF capacitor.

As the target range ($R$) increases, the amplitude of the echo signal decreases. The maximum amplitude of the echo signal as function of the target range is depicted in Fig. 3.6, for the case of a rectangular metal plate and a metal sphere. This graph shows that the maximum echo amplitude can be larger than 1 V for a metal plate at a range of 1 m from the radar and smaller than 2 μV for a metal sphere located at a range of 40 m. The relation between the maximum echo amplitudes of these two targets is around 60 dB.

Although other types of targets can produce echoes with different amplitude values, the radar receiver will be designed to be capable of dealing with at least 60 dB of signal variation and to have an input referred noise level inferior to 10 μV in order to be able to detect echoes from small targets located at a range smaller than 15 m.

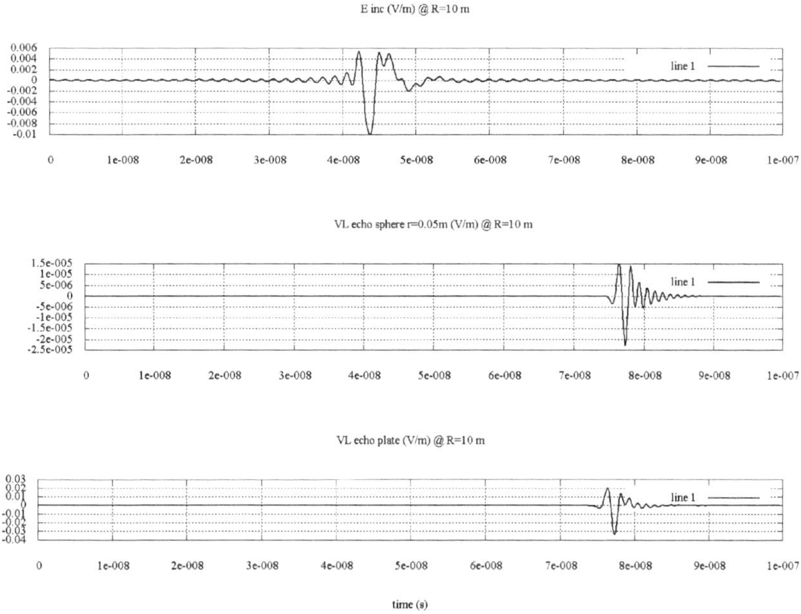


**Fig. 3.5** Received voltages for the echoes from a metal sphere and a rectangular metal plate, together with the incident electrical field at a range equal to 10 m

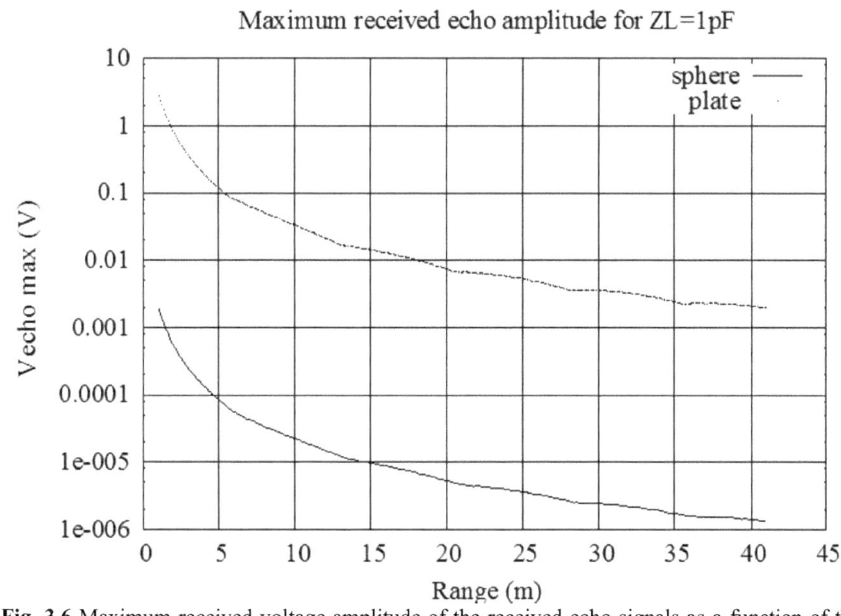


**Fig. 3.6** Maximum received voltage amplitude of the received echo signals as a function of the target range

## 3.3 UWB Receiver Architecture

A receiver architecture for a traditional radar system [1] is based on the super-heterodyne receiver and it is composed by a mixer that converts the received signal to a lower intermediate frequency (IF), the IF frequency value ($f_{IF}$) is equal to $f_{IF}= f_{RF} - f_{LO}$ and must be larger than the pulse bandwidth ($B$). The mixer can be preceded by a low noise amplifier (LNA) to increase the sensitivity of the receiver, but this circuit can also limit the dynamic range of the receiver. The output signal from the mixer (IF signal) is usually amplified using a matched filter to maximize the peak signal to mean noise ratio, the amplitude of the amplified signal is detected using a diode followed by a low pass-filter, this detector type loses the phase information from the signal. The resulting output signal can be displayed as a function of time using an appropriate timing signal whose frequency is related to the *PRF*. A target echo is located where this signal amplitude exceeds a given threshold and the target position is determined by the relation between the display timing signal and the *PRF* signal. A simplified block diagram depicting this architecture is shown in Fig. 3.7.

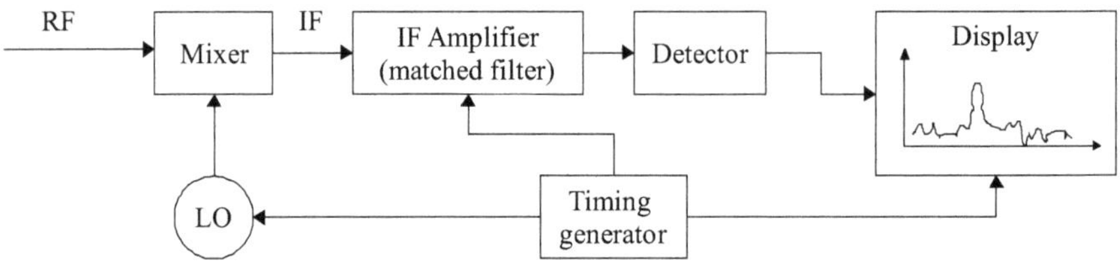


**Fig.** 3.7 Simplified block diagram of traditional radar receiver architecture

More recent receiver radar architectures digitize the signal after the mixer circuit (which produces phase and quadrature outputs) and use digital signal processing techniques to implement the matched filter and the detector. This allows for coherent detection of the pulse where both phase and amplitude information of the input signal are used to perform target detection [1].

### *3.3.1 The Sub-Sampling Technique*

The architecture of an UWB radar receiver can be simpler, since the UWB signal is already at base band, it is not necessary to use a mixer to down-convert the received signal to an IF frequency. The signal can be directly digitized using an ADC. The sampling frequency of the ADC should be at least two times higher than the pulse bandwidth. Considering as an example a Gaussian pulse with 1 ns duration ($\tau$), the sampling rate value of the signal should be at least 2 GHz (Nyquist frequency). Building an ADC capable of working with such a high sampling frequency value is a very difficult task. However, assuming that the targets illuminated by the radiated pulse do not move significantly from pulse to pulse (this is especially true for short range radar systems using high *PRF* values), the received echo signal will remain constant from pulse to pulse (see Appendix A). As an example, a target moving with a speed of 1 m/s and a *PRF* frequency equal to 1 MHz, results that the target moves 1 μm between transmitted pulses, which corresponds to a time difference between the arrivals of two successive echoes of approximately 3 ps. This example shows that the received signal (echo signal) can be considered a periodic signal for slow moving targets and high *PRF* values.

Periodic signals can be sampled using sub-sampling, this technique uses a sample clock with a frequency lower than the required Nyquist frequency, the sampling frequency value is such that each clock period is slightly larger (or smaller) than the period of the *PRF* clock, thus in each sample clock a different point of the received signal will be sampled. The samples obtained in this manner are the same as the ones obtained if the signal was sampled using an equivalent sampling frequency value equal to $1/\Delta T_1$. The resulting sampled signal can be represented in an equivalent expanded time. This process is depicted in Fig. 3.8.

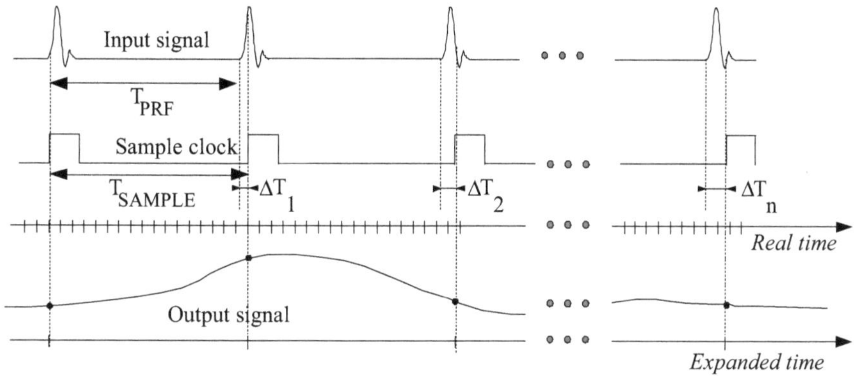


**Fig. 3.8** Sub-sampling of the received signal, this procedure is equivalent to sampling the (repetitive) input signal with a sampling frequency equal to $1/\Delta T_1$

The expanded time signal has the same shape than the original signal since the sub-sampling operation produces the same samples that would be obtained if a regular sampling operation was performed. The sub-sampled signal can be analyzed in order to determine the presence of targets, the simplest way to detect a target consists in comparing the sub-sampled signal with an appropriate threshold value. When the sub-sampled signal exceeds the given reference level, an echo from a target is considered to be present. The exact instant when the echo signal crosses the threshold value can be determined using the relation between the expanded time domain and the real time domain. Noting that the sub-sample operation produces samples that correspond to the input signal value at precise time instants (spaced apart by $\Delta T_1$) and that each one of these time instants has a precise delay to the initial pulse transmission, it is easy to understand that each sampling instant corresponds to a precise distance (or range) from the radar. These sampling instants can be considered range gates. An echo signal can have a duration larger than $\Delta T_1$ and occupy several sample instants, but in that case the distance to the target can be determined using the first range gate where the echo signal exceeded the threshold value.

### 3.3.2 The Problem of Generating the Sampling Clock

The sub-sampling frequency value must be such that the difference ($\Delta T_1$) between the sampling ($T_{SAMPLE}$) period and the *PRF* period ($T_{PRF}$) is equal or smaller than the period of the input signal Nyquist sampling frequency. Considering that the input signal has a bandwidth equal to $B$, the value of the Nyquist frequency is given by

$$F_{S\ Nyquist} = 2 \times B \tag{3.20}$$

The value of the difference between the *PRF* period and the sample period is given by (this value is the equivalent sampling period and should be inferior to the Nyquist period)

$$\Delta T_1 = \left| T_{PRF} - T_{SAMPLE} \right| \leq \frac{1}{F_{S\ Nyquist}} \tag{3.21}$$

The sub-sampling frequency value is given by

$$F_{SAMPLE} = \frac{1}{\left| T_{PRF} - \Delta T_1 \right|} \tag{3.22}$$

As in normal sampling operation, it is desirable to use an equivalent sampling frequency higher than the Nyquist frequency. This means that the difference ($\Delta f$) between the sub-sampling frequency value and the *PRF* value should be small, corresponding to a small $\Delta T_1$ value. These two quantities can be related using

$$\Delta T_1 = \frac{1}{PRF} - \frac{1}{PRF + \Delta f} = \frac{\Delta f}{PRF \cdot \left( PRF + \Delta f \right)} \approx \frac{\Delta f}{PRF^2} \tag{3.23}$$

The maximum value of $\Delta f$ can be obtained combining expressions (3.20) (3.21) and (3.23) and is given by

$$\Delta f = PRF \cdot \frac{PRF}{2 \times B} \tag{3.24}$$

Considering that the input signal has a bandwidth of 1 GHz and that the *PRF* value is equal to 1 MHz, results in a $\Delta f$ value equal to 500 Hz, as previously stated, this corresponds to the maximum value and, in reality, a smaller value should be used. This small value can be a problem, since it is necessary to generate two clock signals with a very small relative difference between their frequencies values (in this case $PRF$ = 1 MHz and $F_{SAMPLE}$ =1.0005 MHz). One of the clock signals (*PRF*) can be obtained directly from a crystal oscillator and the other can be created using a phase lock loop (PLL) [17], as shown in Fig. 3.9.

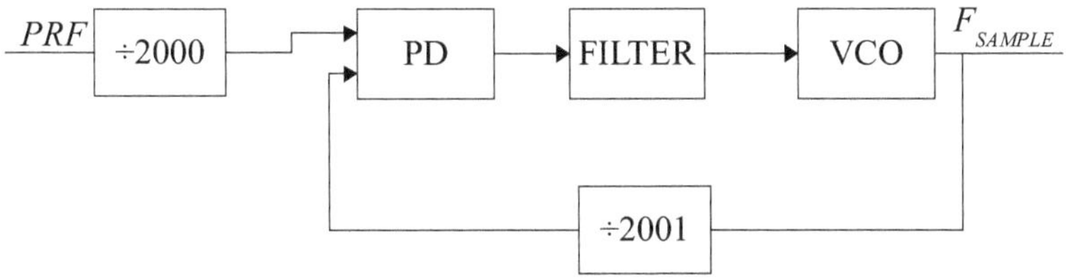


**Fig. 3.9** Generation of the sample clock from the *PRF* clock using a PLL. The coefficients are set so that a PRF frequency of 1 MHz produces a $F_{SAMPLE}$ frequency equal to 1.0005 MHz

The phase detector (PD) of this PLL works at a very low frequency (in the case of the previous example at 500 Hz), the PLL closed loop bandwidth must be, at least, 10 times smaller than the frequency of the PD input signals to guaranty a stable PLL operation. The PLL loop (low-pass) filter must have a low cut-off frequency in order to obtain the desired PLL closed loop bandwidth and consequently needs to occupy a large area (or use discrete components) due to the large capacitance and resistance values needed to obtain large time constant values. If the difference between the *PRF* frequency value and the sample frequency is smaller (as would be desired) the PLL bandwidth would become increasingly small making the filter area even larger and extremely difficult to integrate in CMOS technology. The small loop filter bandwidth would also result in a very long settling time for the PLL.

### *3.3.3 Using a Programmable Delay*

Instead of using two signals with different frequency values, a better approach is to use the same frequency for both the sample clock and the *PRF* clock and create the sample clock signal by introducing a variable delay into the *PRF* clock signal. In this case, the sampling instant is defined by the rising (or falling) edge of the *PRF* clock plus the time delay between the sampling clock and the *PRF* clock. This approach has the advantage of allowing a better control over the sampling operation. Each time delay value defines a distance or range where a target can be located. Using this variable delay, it is possible to perform a non-linear sweep of the radar range. By selecting at which range distances the target might be located and adjusting the variable delay value in order to define the sampling instants around range distances of interest. This allows the radar receiver to skip ranges where there are no targets of interest. This is useful if certain portions of the radar range are to be avoided[5] or if the radar is operated as bubble alarm (where the objective is to determine if a target enters a security bubble). If two separate frequencies are used (one to define the pulse transmission instants and the other to define the sampling instants) the radar receiver is only capable of sweeping the distance range gates sequentially.

If the input signal is sub-sampled at the equivalent Nyquist frequency rate, the time needed to sweep the entire radar range sequentially is given by

$$T_{SWEEP} = T_{PRF} \cdot \frac{T_{PRF}}{T_{NYQUIST}} \tag{3.25}$$

From a signal processing point of view it is better to use an equivalent sampling frequency, larger than the Nyquist frequency, therefore expression (3.25) defines the minimum sweep time. This time will increase if a better range resolution (defined as the number of samples per sweep) is desired. Using a sample clock generated by adding a variable delay to the *PRF* clock, it is possible to increase the number of samples (by reducing the amount of delay between two consecutive samples) around the range where the target is located and reducing (or eliminating) the number of samples elsewhere in range, therefore improving the resolution of the sub-sampled signal while maintaining the same sweep time (same number of samples per sweep).

A simplified version of an UWB receiver architecture using a delayed sample clock is depicted in Fig. 3.10.

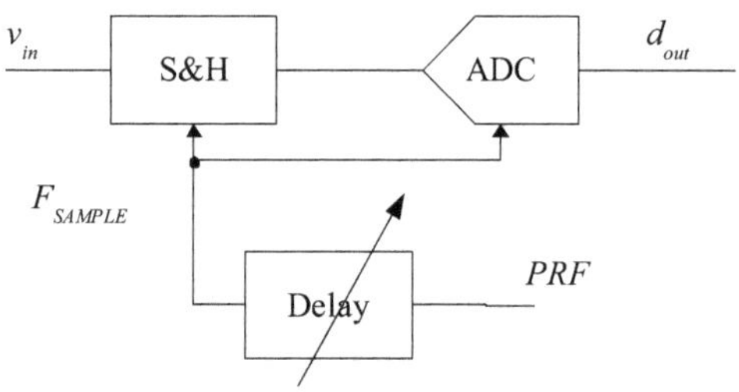


**Fig.** 3.10 UWB receiver architecture using a delayed sample clock

The UWB receiver is constituted by a sample and hold circuit followed by an ADC. Both of these circuits have the same sampling frequency value. The sampling clock ($F_{SAMPLE}$) is a delayed version of the *PRF* clock that controls the time when the transmit pulse is radiated. The amount of delay ($T_D$) between these two clocks is controlled using a variable delay circuit. The maximum delay value ($T_{Dmax}$) should be equal to the delay of an echo produced by a target located at the

[5] This is important for GPR where the radiated pulse has a large power and the echo created by transition from air to earth is very strong and should be avoided.

radar maximum range. The minimum delay value ($T_{Dmin}$) determines closest range where a target can be located. The minimum value of the delay increment ($(\Delta T_D)_{min}$) produced by the variable delay must be inferior to the value $\Delta T_1$ given by expression 3.21 (i.e. the value determined by the equivalent Nyquist frequency). The total number of samples ($N_{SWEEP}$) produced by a complete range sweep is given by

$$N_{SWEEP} = \frac{T_{D\max} - T_{D\min}}{\left(\Delta T_{D\min}\right)_{\min}} \tag{3.26}$$

Considering that the variable delay circuit is digitally programmed, i.e. it is a digital-to-time converter (DTC), the programming word must have $log_2(N_{SWEEP})$ bits in order to obtain $N_{SWEEP}$ of samples per sweep. The minimum possible variation in the programming word (minimum delay programming step or *lsb*) produces a delay variation equal to $(\Delta T_D)_{min}$. This value determines the timing resolution that a return echo can be measured, i.e. it determines the resolution of the target distance measured by the radar. The radar ranging resolution (in meters) is given by

$$\Delta R_{\min} = \frac{c \cdot \left(\Delta T_{D\min}\right)_{\min}}{2} \tag{3.27}$$

If, for example, a ranging resolution of 3 cm is desired, the minimum programming step (*lsb*) of the DTC ($(\Delta T_D)_{min}$) should be inferior to 200 ps. Note that using this value makes the sub-sampling operation equivalent to sampling the input signal with a sampling frequency value equal to 5 GHz.

### 3.3.4 The Sample and Hold

Although the sub-sampling frequency value ($F_{SAMPLE}$ equal to the *PRF* value), is much lower than the signal bandwidth, the input of the sample and hold circuit must still have a input bandwidth larger than the input signal bandwidth $B$ (around 1 GHz for pulses with 1 ns duration) in order to adequately sample the input signal. A simplified sample and hold circuit (S&H) is a switch controlled by the sample clock and a capacitor (shown in Fig. 3.11- left side). A more realistic S&H circuit would need an operational amplifier to isolate the capacitor voltage from the circuit connected to the output of the S&H circuit. During the sampling phase, this S&H circuit, can be analyzed as a simple RC circuit (shown in Fig. 3.11- right side) that must have a a bandwidth larger than $B$ in order not to distort significantly the input signal.

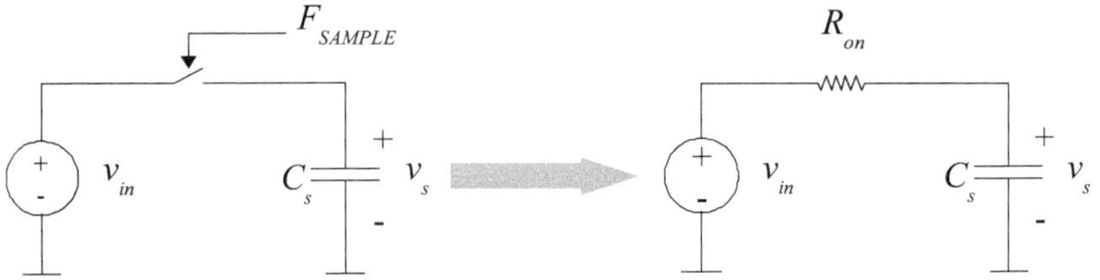


**Fig. 3.11** Simplified Sample & Hold circuit (left side) and simplified Sample & Hold circuit during the sampling phase (right side)

The bandwidth of the circuit is determined by the inverse of the $R_{on}.C_s$ time constant, where $R_{on}$ is the resistance value of the switch when it is closed and $C_s$ is the sampling capacitance value. Increasing the bandwidth of the circuit translates into reducing the values of $C_s$ and $R_{on}$. However the value of $C_s$ determines the power of the noise stored in the capacitor at the end of the sampling phase. This noise results from the ON resistance of the switch, this resistance creates thermal noise which can be represented by a voltage source with a PSD equal to $4 \cdot K_B \cdot T \cdot R_{ON}$. This PSD is filtered by the low-pass circuit formed by the switch resistance and sampling capacitor. Since this noise voltage is also sampled by the capacitor, its PSD is folded into the frequency band ranging from 0 Hz to $F_{SAMPLE}/2$. The sampled noise power is calculated using [18]

$$V_N^2 = \int_{f=0}^{\infty} 4 \cdot K_B \cdot T \cdot R_{ON} \cdot \left| \frac{1}{j \cdot 2 \cdot \pi \cdot f \cdot R_{ON} \cdot C_S + 1} \right|^2 df = \frac{K_B \cdot T}{C_S} \tag{3.28}$$

where $K_B$ is the Boltzmann constant and $T$ is the temperature in Kelvin. From expression 3.28 it is clear that the value of $C_s$ should be larger than a certain value to obtain a noise power below a desired level, this means that the input impedance of the circuit in Fig. 3.11 decreases as the desired noise level is lowered (because the value of $C_s$ increases). The value of $R_{on}$ must also decrease as the value of $C_s$ increases in order to maintain the desired input bandwidth for the circuit. If the circuit is designed to have an input bandwidth equal to $B$ and a noise level equal to $V_N^2$, the circuit input impedance, for a sine-wave input signal with a frequency value equal to $B$, is given by

$$Z_{in}(B, V_N^2) = R_{ON} + \frac{1}{j \cdot 2 \cdot \pi \cdot B \cdot C_S} = \frac{V_N^2}{K_B \cdot T} \cdot \frac{1}{2 \cdot \pi \cdot B} \cdot (1 - j) \tag{3.29}$$

The input impedance value is inversely proportional to the desired bandwidth value ($B$) and proportional to the desired noise level ($V_N^2$). This expression shows that the circuit of Fig. 3.11 can not be designed to have a bandwidth ($B$) and noise level ($V_N^2$) independent from the input impedance value ($Z_{in}$). If the desired noise level is very small or the desired input bandwidth is very large, the resulting S&H

input impedance value ($Z_{in}$) can be very small. The equivalent model of the antenna (or any signal source) is a Thevenin equivalent circuit, composed by a voltage source ($V_S$) in series with an impedance ($Z_S$). Therefore the amplitude of a sampled sine-wave, with frequency $\omega$, is given by

$$V_A(\omega) = V_{in}(\omega) \cdot \frac{Z_{in}(\omega, V_N^2)}{Z_{in}(\omega, V_N^2) + Z_S(\omega)} \tag{3.30}$$

The previous expression shows that if $Z_{in}$ becomes much smaller than $Z_s$, a significant amount of attenuation can be introduced into the input signal. This means that although the noise power is constant, i.e. does not depend on the input signal frequency, the Signal-to-Noise-Ratio (*SNR*) of the sampled sine-wave at the output of the S&H depends on the frequency of the input signal.

Traditionally the noise performance of RF systems is evaluated using the Noise Figure (*NF*), which is defined as the *SNR* of the input signal divided by the *SNR* at the output of the system.

$$NF(f) = \frac{SNR_{input}}{SNR_{output}} = \frac{S_S(f)}{S_{N\ input}(f)} \cdot \frac{S_{N\ circuit}(f) + G(f) \cdot S_{N\ input}(f)}{G(f) \cdot S_S(f)} \tag{3.31}$$

where $G(f)$ is the power gain of the system (the squared result of dividing expression 3.30 by $V_{in}$), $S_S(f)$ is the PSD of the input signal, $S_{N\ input}(f)$ is the PSD of the input noise and $S_{N\ circuit}(f)$ is the PSD of the noise created by the circuit. Using expression 3.31 to calculate the *NF*, results that in a given system, the *NF* value depends on the frequency value, this is why this definition is also know as the spot *NF*. However, this definition of the *NF* is not suited for UWB signals, since the spot *NF* value can change a lot within the bandwidth of the UWB signal and therefore does not provide a unique value to characterize the performance of an UWB system. In the case of narrow band systems, the input signal can be considered a sine-wave and the signal power is simply the amplitude of the input sine-wave squared. The system power gain and the noise PSD can be considered constant within the signal band, resulting that in the case of narrow band systems the spot *NF* expression can be simplified to

$$NF(f_0) = 1 + \frac{S_{N\ circuit}(f_0) \cdot B}{G(f_0) \cdot S_{N\ input}(f_0) \cdot B} \tag{3.32}$$

where $B$ is the signal (narrow) band and $f_0$ is signal central frequency. This expression is similar to another *NF* definition the power *NF*, which uses the signal and noise powers integrated over the desired bandwidth, instead of the signal and noise PSD. The power *NF* does not depend on the signal frequency as the spot *NF*, but provides a performance indication for the system performance that might not be related to the desired system performance [19]. In the case of the S&H circuit, the most important parameter is the value of the output amplitude at a given time instant (desired sampled voltage) and not the value of the average output power over a given time interval. A better indicator of the S&H noise performance is obtained defining the output *SNR* as the ratio of the maximum sampled voltage value (the output S&H voltage at the end of the sampling phase), divided by the total noise *rms* voltage present in the S&H output voltage, given by expression 3.28 (this expression includes both the noise of the circuit, due to $R_{on}$, and the noise of the input signal, due to $R_S$).

In order to obtain a simple expression for the maximum output amplitude of the S&H output voltage for an UWB input signal, it is convenient to use a rectangular pulse (see Section 2.2) as the input UWB signal. This type of signal simplifies the calculation of the circuit's voltages as a function of time, if a Gaussian pulse is used instead, it is not possible to obtain an analytical expression for the circuit's output voltage. A rectangular pulse can be mathematically defined as the sum of two step functions, where the second step function has a negative amplitude and a time delay equal to the pulse duration. It is also convenient, in order to simplify the calculations, to consider that the antenna impedance is real and constant over frequency, with a value equal to $R_S$. The antenna impedance is a complex value that varies with the signal frequency. However, if a well designed wide-band antenna is used, such as a conical or horn antenna (see Section 2.3), the antenna exhibits a mostly real constant impedance within the signal band. The output voltage of the circuit in Fig. 3.11 for a rectangular pulse input signal corresponds to the step response of the circuit. The sub-sampling operation produces a signal with the shape of $v_s(t)$ but expanded in time. The sampled voltage value depends on the sampling instant time. The maximum value of the sampled voltage (peak voltage) for a rectangular pulse input signal occurs when the sampling instant of the S&H circuit coincides with the falling edge of the rectangular pulse (this condition can be obtained by adjusting the delay value of the sample clock). Under these conditions, the maximum sampled voltage value for an input rectangular pulse with duration $\tau$ and amplitude $A$, occurs when $t=\tau$ and it is given by

$$v_{sample} = v_S(\tau) = A \cdot \left[ 1 - e^{-\frac{\tau}{(R_S + R_{on}) \cdot C_S}} \right] \tag{3.33}$$

This is illustrated in Fig. 3.12. The $v_s(t)$ signal is very different from the input signal because the rectangular pulse has a large bandwidth, the abrupt edges of the this pulse require an almost infinite bandwidth in order to have a reasonable rectangular shape. However, the desired information is contained only in the pulse amplitude.

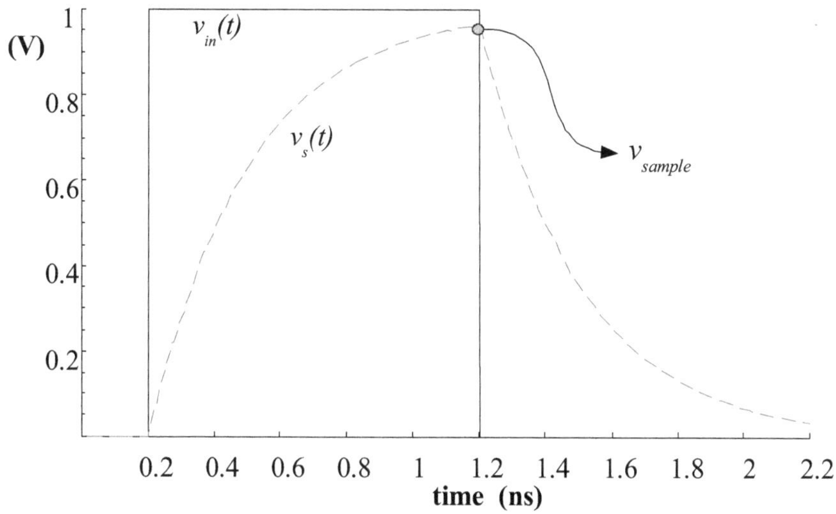


**Fig. 3.12** Step response of the S&H circuit ($v_s(t)$) for an input rectangular pulse ($v_{in}(t)$) with $A$= 1 V, τ=1 ns, $R_{on}$=100 Ω and $C_s$=2 pF . The maximum sampled voltage (peak echo amplitude) occurs when the sampling time instant coincides with the end of the pulse. The sampled voltage is smaller than A because the S&H circuit does not have enough time to settle

This expression is used to calculate the *SNR* of maximum echo amplitude at the output of the S&H, although the expression results from a simplified analysis, it can be used to compare the performance of different type of circuits and different circuit's designs.

$$SNR_{dB} = 20 \times \log \left[ \frac{A \cdot \left[ 1 - e^{-\frac{\tau}{(R_S + R_{on}) \cdot C_S}} \right]}{\sqrt{\frac{K_B \cdot T}{C_S}}} \right] \tag{3.34}$$

The *SNR* value is a function of the sampling capacitance value, the ON resistance value of the switch and the antenna impedance. For an input rectangular signal with an amplitude $A$ equal to 1 V, a duration $\tau$ equal to 1 ns and a $R_S$ value equal to 50 Ω, the *SNR* as function of $C_S$ and three different values of $R_{ON}$ is plotted in Fig. 3.13. The temperature is equal to 290 Kelvin This graph shows that, as expected, the *SNR* increases with the increase of the sampling capacitance value, but if this capacitance value increases beyond a certain value, the $SNR_{out}$ will decrease (due to the limitation of the circuit's bandwidth). The maximum $SNR_{out}$ value depends on the bandwidth value of the S&H circuit, defined by the $R_{ON}$, $R_s$ and $C_s$ values. The graph also shows that the S&H switch must have a very low value for its ON resistance (below 100 Ω) in order to adequately sample the UWB signal.

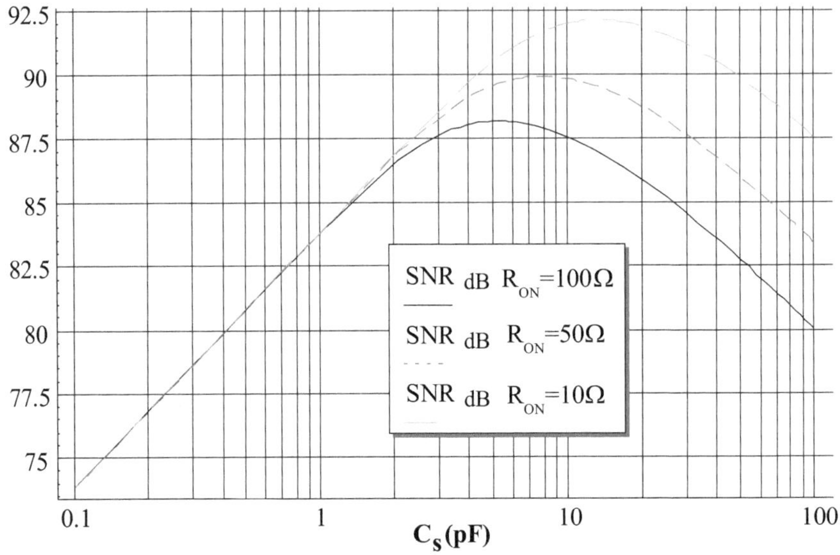


**Fig. 3.13** $SNR_{out}$ of the maximum sampled voltage value as a function the sampling capacitance value and different ON resistance values of the switch. The rectangular input pulse has $A$= 1 V and $\tau$=1 ns

The previous graph shows that the *SNR* increases with increasing values of $C_s$ because the output noise power is progressively reduced, but after a certain $C_s$ value the attenuation of the input signal starts to increase significantly (due to the low value of the capacitor impedance) and hence the *SNR* achieves its maximum value and starts to decrease. The maximum *SNR* value depends on the $R_{ON}$ resistance, for $R_{ON}$=50 Ω the maximum *SNR* value at the output of the S&H is obtained using a sampling capacitor value of 10 pF and is equal to 92 dB for an input signal with 1 V amplitude. If the amplitude of the input signal is reduced to 10 μV the expected maximum *SNR* would be reduced to –8 dB. In this case it would not be possible to detect an echo because the maximum echo amplitude value would be well below the *rms* noise level and would be indistinguishable from the noise. From this example it is easy to understand that the *SNR* should be larger than 0 dB in order to detect any echo signal present. The amplitude of the input signal that results in a 0 dB *SNR* is the input referred noise of the S&H circuit, in this case the maximum *SNR* value of 92 dB corresponds to an input referred noise value of –92 dBV (~25 μVrms). The input referred noise is an important parameter in the analysis of the radar receiver performance. The input referred noise can be compared against the expected received peak amplitude of the echo signal in order to determine the radar maximum range. The input referred noise value is given by

$$V_{Nin\ SH}^{2}=\frac{K_B\cdot T}{C_S}\cdot\frac{1}{\left(1-e^{-\tau/(R_S+R_{on})\cdot C_S}\right)^{2}}\tag{3.35}$$

The previous analysis showed that in a S&H circuit it is not possible to improve the output SNR indefinitely. A minimum value for the input referred noise is obtained for an optimum sampling capacitance value. This value depends on the source impedance value ($R_s$) and on the pulse width. In order to continue to reduce the input referred noise of the receiver beyond this optimum value other options must be considered.

### 3.3.5 Averaging Samples Together

One of the simplest solutions to improve the *SNR* of the output of the S&H circuit is to calculate the average of a set of (equal) output samples. This averaging operation consists in simply adding the samples together. This solution has been used in the past, in sampling oscilloscopes [20] and in the UWB low cost radar system described in [21]. This technique imposes the restriction that the echo signal can not change significantly during the time it takes to add the required number ($n$) of samples together. Each sample is composed by the echo signal for a given time delay value ($T_0$) and a random value corresponding to the noise voltage ($V_N(t)$), the resulting equivalent sample voltage value is given by

$$V_{equivalent}=\sum_{i=1}^{n}v_{sample}\left[i\right]=\sum_{i=1}^{n}\left(V_{echo}\left(T_0\right)+V_N\left(t_i\right)\right)=n\cdot V_{echo}\left(T_0\right)+\sum_{i=1}^{n}V_N\left(t_i\right)\tag{3.36}$$

When $n$ samples of the echo signal are added together, the amplitude of the echo signal increases $n$ times resulting in an increase in the signal power of $n^2$ times. Adding random numbers together is equivalent to performing the averaging operation. The result of adding $n$ random numbers with a variance $\sigma^2$ is a random number with a variance $n.\sigma^2$ [18]. Therefore the power of the noise increases only $n$ times. The *SNR* of the average signal can be calculated using

$$SNR_{average}=10\times\log\left[\frac{\left(n\cdot v_{echo}\right)^{2}}{n\cdot V_N^2}\right]=10\times\log\left(n\right)+SNR_{dB}\tag{3.37}$$

where $SNR_{dB}$ is given by expression 3.34. Expression 3.37 shows that for each doubling of the number of samples ($n$) added together the resulting SNR will improve by 3 dB. As an example if 1000 samples are added together, the SNR would improve by 30 dB. Of course, improving the *SNR* using this method increases the time necessary to perform a range sweep ($T_{SWEEP}$) by $n$ times. It requires that the target does not move significantly during the time necessary to sample its echo signal, but even a target moving at 1 m/s would only move 1 mm during the time it takes to average 1000 echoes with a *PRF* of 1 MHz.

A simple circuit that implements the function of an averaging S&H is shown in Fig. 3.14. Note that this circuit is essentially the same as a simple Switched-Capacitor S&H, the only difference is the timing of the clock phases (specifically the reset phase). Therefore the averaging operation does not introduce a power dissipation penalty into the radar system.[6]

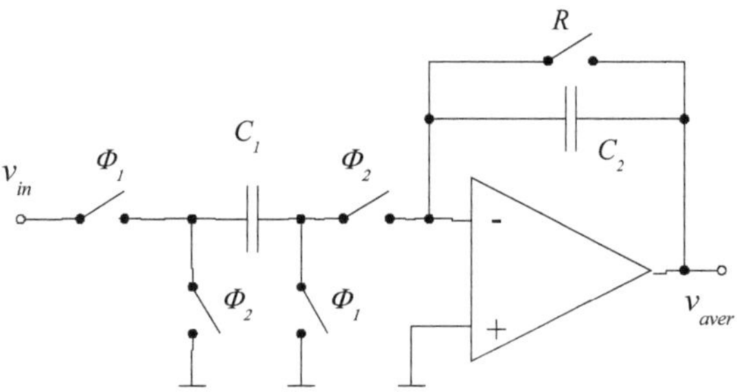


**Fig. 3.14** Averaging S&H circuit. The sampling instant is determined by the falling edge of phase $\Phi_1$. Before a new average sample value is produced, it is necessary to reset capacitor $C_2$

The clock phases $\Phi_1$ and $\Phi_2$ are non-overlapping and are generated from the delayed clock (sample clock). The operation of the circuit is as follows first, the feedback capacitor is reseted, then capacitor $C_1$ is charged by the input voltage and at the falling edge of the sampling clock (phase $\Phi_1$) stores a charge proportional to the input voltage value. During phase $\Phi_2$ this charge is transferred into the feedback capacitor ($C_2$) producing a voltage equal to $V_o = {}^{C_1}\!/_{C_2} \cdot V_{sample}$ ($V_{sample}$ is given by expression 3.33). This process is repeated during the next $n$ clock cycles, resulting in an output voltage, at the end of $n$ clock cycle, equal to

$$V_o = \frac{C_1}{C_2}\sum_{i=1}^{n}\left(V_{sample}\left(T_0\right)+V_N\left(t_i\right)\right) \tag{3.38}$$

This expression shows that the signal ($V_{sample}(T_0)$) is amplified by ${}^{C_1}\!/_{C_2} \times n$ if $n$ samples are added together, but adding $n$ noise samples ($V_N(t_i)$) together will only

[6] Of course a more power efficient S&H circuit would use the same capacitor to sample the input voltage during $\Phi_1$ and then connected it as the feedback capacitor during $\Phi_2$. This would result in circuit with a feedback factor close to 1 and hence the amplifier could be designed to have a lower GBW value, corresponding to a lower power dissipation. But for a sampling frequency value around 1 MHz the power penalty of the depicted circuit is very small.

increase the variance by $n$ times. Applying the Z transform to equation 3.38 it is possible to obtain the circuit transfer function

$$H(z)=\frac{V_o(z)}{V_{in}(z)}=\frac{C_1}{C_2}\cdot\sum_{i=0}^{n-1}z^{-i}=\frac{C_1}{C_2}\cdot\frac{1-z^{-n}}{1-z^{-1}} \tag{3.39}$$

This transfer function corresponds to a discrete *sinc* filter. Since the input sample has always the same value ($V_{sample}(T_0)$) it can be considered a DC signal ($z$=1), note that $H(z=1)={C_1}/{C_2}\times n$. Any noise signal present in the input will be filtered by the *sinc* function, the noise present at some frequency bands will be eliminated (the *sinc* function has zeros located at frequencies values equal to the multiples of $F_{SAMPLE}/n$) and the noise present at other frequency bands will be attenuated by the circuit.

The total noise power at the output of the circuit is due to several noise sources [22]. When capacitor $C_1$ samples the input voltage, the ON resistance of the switches causes a noise voltage to appear in capacitor $C_1$ at the end of the sampling phase. The power of this noise voltage is given by expression 3.28, during phase $\Phi_2$ the charge corresponding to this voltage is transferred to the feedback capacitor $C_2$. During phase $\Phi_2$ the capacitor $C_1$ is connected to the virtual ground of the amplifier through switches that also have a ON resistance and therefore also add noise to the output voltage. The power of this noise voltage is also given by expression 3.28. In each clock cycle a different value of noise power (corresponding to the noise voltage at time instant $t_i$ $V_N(t_i)$) is added together. At the end of the $n$ clock cycles the total noise power present in the output voltage will be $n$ times larger than the noise power added in each clock cycle to the output voltage.

The amplifier also contributes with noise power to the output voltage. The amplifier noise is independent of the number of samples added together since the amplifier is always connected to the output node; therefore its output noise power is always constant and only depends on the amplifier bandwidth. Therefore, the total noise power at the output of the circuit of Fig. 3.14, after $n$ samples are added together, is given by

$$V_{No}^2=\frac{K_B\cdot T}{C_2}+\left(\frac{C_1+C_2}{C_2}\right)^2\cdot V_{Nopamp}^2+\left(\frac{C_1}{C_2}\right)^2\cdot\sum_{i=1}^{n}\left(2\times\frac{K_B\cdot T}{C_1}\right) \tag{3.40}$$

The contribution of capacitor $C_2$ to the total noise power is due to the reset switch, activated once during the reset phase. The input referred noise of the circuit can be calculated dividing expression 3.40 by the signal gain squared, resulting in

$$V_{N\,in}^2 = \left[\frac{1}{n^2}\cdot\left(\frac{C_2}{C_1}\right)^2\cdot\left(\frac{K_B\cdot T}{C_2}+\left(\frac{C_1+C_2}{C_2}\right)^2\cdot V_{Nopamp}^2\right)+\frac{2}{n}\times\frac{K_B\cdot T}{C_1}\right]\cdot\frac{1}{\left(1-e^{-\tau/((R_{on}+R_S)\cdot C_1)}\right)^2} \tag{3.41}$$

The previous expression shows that the contributions of the amplifier noise and reset switch noise to the input referred noise are not very important since they are divided by $n^2$. The input referred noise of the circuit is shown in Fig. 3.15 as a function of the sampling capacitor ($C_1$). The series resistance of the sampling switches is 50 Ω, the amplifier noise is 100 μVrms and the number of samples added ($n$) is equal to 100. The temperature in this analysis is equal to 290 Kelvin.

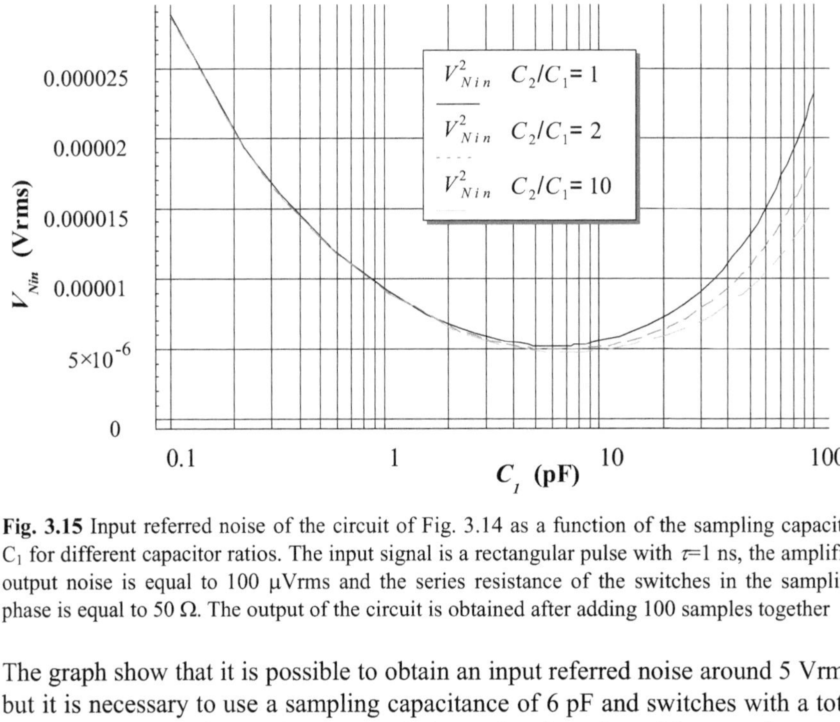


**Fig. 3.15** Input referred noise of the circuit of Fig. 3.14 as a function of the sampling capacitor $C_1$ for different capacitor ratios. The input signal is a rectangular pulse with $\tau$=1 ns, the amplifier output noise is equal to 100 μVrms and the series resistance of the switches in the sampling phase is equal to 50 Ω. The output of the circuit is obtained after adding 100 samples together

The graph show that it is possible to obtain an input referred noise around 5 Vrms, but it is necessary to use a sampling capacitance of 6 pF and switches with a total ON resistance equal to 50 Ω. The graph shows that the value of the capacitance ratio ($C_2/C_1$) is not important in determining the value of the optimum input referred noise of the circuit. This is because of the large value of $n$, which reduces the noise contribution from the amplifier and the reset switch.

There is another advantage of adding $n$ samples together, since the voltage gain is given by $C_1/C_2 \times n$. It is very easy to control the amount of desired gain simply by changing the number of samples added together. This means that implementing an automatic gain control (AGC) function at system level is very easy. An example of where the AGC function is useful in the radar system, is when performing a range sweep, it is convenient to increase the value of $n$ as more distant ranges are probed for the echoes of possible targets.

As shown in Appendix B, obtaining switches with the required small value of the ON resistance to sample the input signal with a small attenuation results in large size MOS transistors. These devices have large parasitic capacitors that introduce attenuation in the signal path. Due to this attenuation, the input referred noise values shown in the graph of Fig. 3.15 are smaller than what would be possible in a real CMOS circuit. In order to adequately sample the short duration UWB signal it is necessary to use very large MOS transistors (with channel widths larger than 1 mm) that can introduce a large amount of clock feed-through and charge injection into the sampling capacitor. Some of the charge injection and clock feed-through problems can be mitigated by using a differential circuit implementation. However, due to inevitable mismatches between the transistors, (and the transistors parasitic capacitances), part of the charge injection and clock feed-trough results in an offset voltage. The amplifier will also have an offset voltage, due to mismatch errors between its constituting transistors. These offset voltages will be added to the output voltage each time an input sample is added to the output voltage. This means that the output of the circuit can exceed the amplifier output voltage swing, before the required number of samples is added together necessary to obtain the desired input referred noise. Although in theory, it would be possible to increase indefinitely the number of averages, assuming that the target is not moving (or moving very slowly), any real S&H circuit will have offset. Since this offset will also be added $n$ times, after a certain value of $n$ the output of the circuit will saturate, thus setting a limit to the maximum number of samples that can be averaged together. If the maximum output voltage swing of the amplifier is given by $\Delta V_{outmax}$ and the circuit total offset is given by $V_{off}$, the maximum number of samples that can be added together is given by

$$n_{\max} = \frac{\Delta V_{out\max}}{V_{off}} \tag{3.42}$$

As an example, if the maximum voltage swing is 1 V and the total offset is 10 mV, the maximum number of averages would be 100.

## 3.3.6 Using a Low Noise Amplifier

Another common solution to improve the output $SNR$ consists in introducing an amplifier between the antenna and the S&H circuit. By placing an amplifier with a voltage gain equal to $G_1$ before the S&H circuit, the voltage signal at the input of this circuit will be $G_1$ times larger and since the S&H noise power remains constant, the result is that the input referred S&H noise voltage of the new circuit should be reduced by $G_1$. Since the *SNR* at the S&H output is given by the signal power divided by the total noise power, composed of the S&H noise power and of the noise power due to the antenna resistance, the minimum possible value for the output *SNR* is the value of the input *SNR*. This value would be achieved if the amplifier had an infinite voltage gain.

The amplifier can be designed to have an input impedance that allows for maximum power transfer from the antenna, or to have an input impedance much higher than the antenna impedance to provide maximum voltage gain [23]. In the first case, the amplifier input impedance is equal to the conjugate value of the antenna impedance, resulting that the amplifier input voltage will be half of the antenna open-circuit voltage, which corresponds to a –6 dB voltage attenuation of the antenna open-circuit voltage. In the second case, since the amplifier input impedance is much larger than the antenna impedance, the amplifier input voltage is equal to the antenna open circuit voltage. If the LNA input impedance is not matched to the antenna impedance, there will be multiple reflections of the input signal that can create multiple echoes. But, if the physical distance between the antenna and the LNA is smaller than the pulse width ($\tau.c$), these reflections will not be confused with echoes from possible targets. The amplifier output impedance should be much smaller than the S&H input impedance (given by expression 3.29), over the desired signal bandwidth in order to avoid signal attenuation between the LNA and S&H circuits.

The amplifier circuit will, inevitably, add noise to the signal path, resulting in a degradation of the output S&H *SNR*. Therefore the amplifier must be designed to have a low noise level, hence it is normally known as a low noise amplifier (LNA). In order to better understand the signals present in the receiver channel path using a LNA, consider the following signal flow diagram (Fig. 3.16)

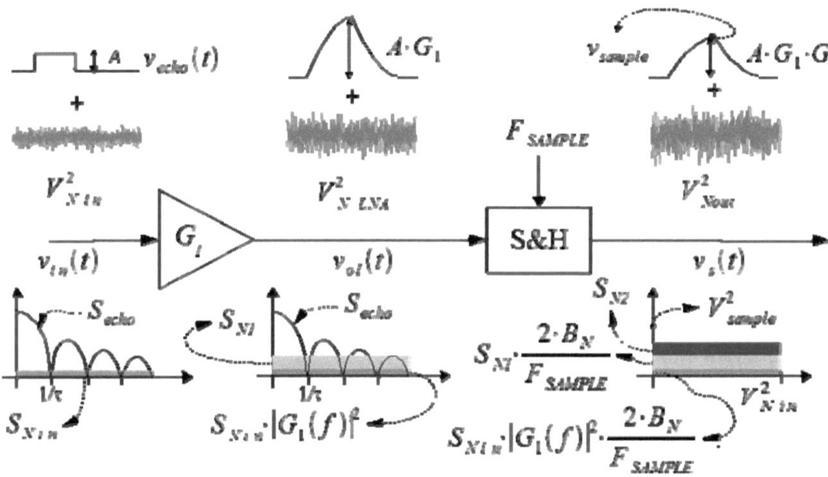


**Fig. 3.16** Signal flow diagram of the receiver. The signals and the noise are shown for each stage; the time domain representation is on the top and the frequency domain on the bottom of the figure

The input signal ($v_{in}(t)$) consists of the desired echo signal ($v_{echo}(t)$) plus some noise received by the antenna ($V_N^2$), at the output of the LNA, the signal ($v_{o1}(t)$) consists of the amplified (and filtered) echo signal, plus the input noise signal, plus some extra noise created by the LNA ($V_{N1}^2$). The S&H circuit samples the LNA output voltage ($v_{o1}(t)$) and produces an output sample ($v_{sample}$), which is constituted by the desired echo signal and by a noise component. The value of this voltage depends on the desired echo signal amplitude, the input noise power (amplified and filtered by the LNA and S&H circuits), the LNA noise (amplified and filtered by the S&H circuit) and the noise due to the S&H circuit itself ($V_{N2}^2$). The total noise power present in the output sample ($v_{sample}$) is larger than the total noise power present in the input signal ($v_{in}(t)$). However the echo signal is also larger because it is amplified by $G_1.G_2$ . The echo signal (in this case a rectangular pulse) has PSD with most of the signal power located in the lower frequencies (see Section 2.2), the noise has an uniform PSD. After the sampling operation, the power of the echo signal (a DC value) is located in a single frequency (0 Hz). The sampled noise PSD is uniform, but now it is distributed from 0 Hz to $F_{SAMPLE}/2$ due to the spectral folding of the sampling operation.

The contribution of the echo signal to the output sample can be calculated considering that the LNA has a first order low-pass transfer function (with a bandwidth equal to $p_1$) given by

$$G_1(s) = \frac{G_0 \cdot p_1}{s + p_1} \tag{3.43}$$

the output voltage of the LNA for a step input with amplitude $A$ is given by

$$v_{o1}(t) = A \cdot G_0 \cdot \left(1 - e^{-p_1 \cdot t}\right) \tag{3.44}$$

The S&H output voltage (as a function of time) can be calculated considering that signal 3.44 is applied to the circuit of Fig. 3.11 (assuming that the LNA output impedance is much smaller than the value of $R_{on}$ in the S&H circuit), resulting in

$$v_s(t) = A \cdot G_0 \cdot \left[1 - \frac{e^{-p_1 \cdot t}}{C_S \cdot p_1 \cdot R_{on} - 1} + \frac{C_S \cdot p_1 \cdot R_{on} \cdot e^{\frac{-t}{C_S \cdot R_{on}}}}{C_S \cdot p_1 \cdot R_{on} - 1}\right] \tag{3.45}$$

The $v_{o1}(t)$ and $v_S(t)$ voltages are plotted in Fig. 3.17, for an input rectangular pulse ($v_{in}(t)$) with $A$=1 V and $\tau$=1 ns and a LNA with a gain[7] $G_0$=10 and

[7] This gain $G_0$ already includes any attenuation due to the mismatch between the antenna impedance and the LNA input impedance.

$p_1 = 2 \times \pi \times 1$ GHz connected to the S&H circuit of Fig. 3.11. The maximum sampled voltage (peak echo amplitude) occurs when the delay in the sample clock is slightly larger than $\tau$, but if $p_1 >> {}^1\!/_{C_S \cdot R_{on}}$ is valid, the ideal sampling instant can be approximated by $\tau$, resulting in a small error in the maximum sampled voltage value. Under this condition the $v_s(t)$ voltage would be a rectangular pulse and the $v_{o1}(t)$ voltage would be given by expression 3.33.

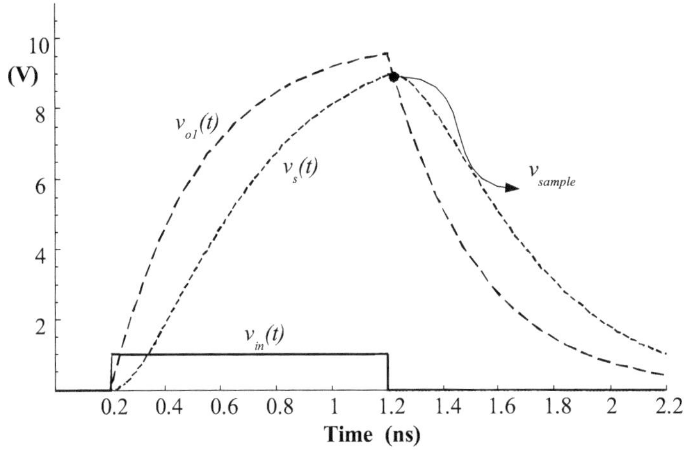


**Fig. 3.17** Step response of the LNA and S&H circuit ($v_{o1}(t)$ and $v_s(t)$) for an input rectangular pulse ($v_{in}(t)$) with $A$= 1 V, $\tau$=1 ns $R_{on}$=100 Ω, $C_s$=2 pF and a LNA with $G_1$=10 and $p_1 = 2 \times \pi \times 1$ GHz. The maximum sampled voltage occurs for a sampling time instant slightly after the end of the input pulse

The peak signal voltage gains ($G_1$ and $G_2$) shown in Fig. 3.17 can be calculated substituting $t$ by $\tau$ in expressions 3.44 and 3.45, respectively and dividing the result by $A$. Assuming that the bandwidth of the LNA is much larger than the bandwidth of the S&H circuit $\left( p_1 >> {}^1\!/_{C_S \cdot R_{on}} \right)$, these voltage gains can be simplified to

$$G_1 \approx G_0 \cdot \left| 1 - e^{-\tau \cdot p_1} \right| \approx G_0$$
$$G_1 \approx G_0 \cdot \left| 1 - e^{\frac{-\tau}{C_S \cdot R_{on}}} \right| \qquad (3.46)$$

Note that the approximate LNA gain value ($G_1$) is the same as it would be for a sine-wave signal with a frequency value smaller than $p_1$, but that the approximate S&H gain value ($G_2$) is different from what it would be for a sine-wave signal with a frequency value inside the S&H bandwidth, which would be approximately equal to 1.

Assuming that all the noise sources in the signal path are white i.e. the noise has an uniform PSD, the PSD of the noise at the output of the LNA can be approximated by

$$S_{N\,LNA}(f) = S_{N1} + |G_1(f)|^2 \cdot S_{Nin} \approx S_{N1} + G_0^2 \cdot S_{Nin} \qquad (3.47)$$

inside the LNA bandwidth ($f < p_1$), where $S_{N1}$ is the PSD of the noise generated by the LNA and $S_{Nin}$ is the PSD of the noise received by the antenna. The spot *NF* of the LNA ($f < p_1$) is defined as

$$NF_{LNA}(f) = \frac{G_0^2 \cdot S_{Nin} + S_{N1}}{G_0^2 \cdot S_{Nin}} = 1 + \frac{S_{N1}}{G_0^2 \cdot S_{Nin}} \qquad (3.48)$$

The total noise power value at the output of the S&H can be calculated assuming that the bandwidth of the LNA is much larger than the bandwidth of the S&H and that the S&H circuit has a noise bandwidth given by [18], [24].

$$B_N = \frac{1}{4 \cdot R_{on} \cdot C_S} \qquad (3.49)$$

The total noise power present is composed by the contributions of the three noise sources in the signal path: input noise from the antenna, LNA noise and S&H noise, therefore

$$V_{Nout}^2 = B_N \cdot \left(G_0^2 \cdot S_{Nin} + S_{N1} + S_{N2}\right) = \frac{K_B \cdot T}{C_S} + \frac{G_0^2 \cdot S_{Nin} + S_{N1}}{4 \cdot R_{on} \cdot C_S} \qquad (3.50)$$

where $S_{N2} = 4 \cdot K_B \cdot T \cdot R_{on}$ is the PSD of the noise created by the $R_{on}$ resistance inside the S&H. Combining Expression 3.50 with expression 3.48 and dividing the result by the LNA and S&H combined gains squared (expression 3.46) results in the input referred noise expression

$$V_{Nin}^2 = B_N \cdot \left(G_0^2 \cdot S_{Nin} + S_{N1} + S_{N2}\right) =$$
$$= \frac{K_B \cdot T}{C_S} \cdot \left(\frac{1}{G_0^2} + \frac{NF_{LNA} \cdot S_{Nin}}{4 \cdot K_B \cdot T \cdot R_{on}}\right) \cdot \frac{1}{\left(1 - e^{-\tau/(R_{on} \cdot C_S)}\right)^2} \qquad (3.51)$$

The input noise voltage PSD received by the antenna ($S_{Nin}$) depends on the ambient temperature ($T_{amb}$, which is not necessarily equal to the circuit operating temperature $T$) of where the antenna beam is aimed and is given by $S_{Nin} = 4 \cdot K_B \cdot T_{amb} \cdot R_a$, where $R_a$ is the real part of the antenna impedance. Therefore the input referred noise power expression can be further simplified to

$$V_{Nin}^2 = \frac{K_B \cdot T}{C_S} \cdot \left( \frac{1}{G_0^2} + NF_{LNA} \cdot \frac{T_{amb} \cdot R_a}{T \cdot R_{on}} \right) \cdot \frac{1}{\left(1 - e^{-\tau/(R_{on} \cdot C_S)}\right)^2} \tag{3.52}$$

This expression was obtained assuming that $NF_{LNA}$ is constant for frequencies below $B_N$, otherwise it would be necessary to integrate the PSD of the noise signals (together with the LNA's $NF$) across the frequency band. The input referred noise when using a LNA with $G_0$=10 for different values is plotted in Fig. 3.18. This graph was obtained for a LNA with a voltage gain of 10 and 3 different values of $NF_{LNA}$ (1, 2 and 3). The $R_a$ value is the same as the $R_{on}$ which is equal to 50 Ω. The temperature value is equal to 290 Kelvin.

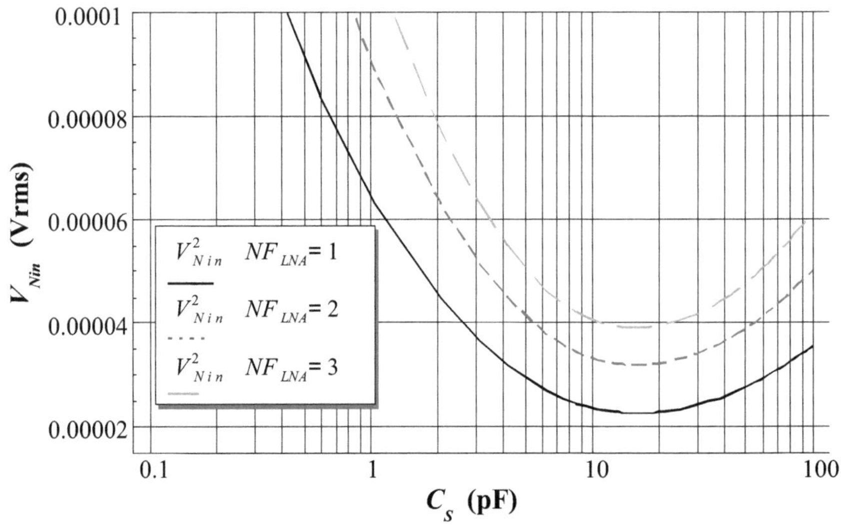


**Fig. 3.18** Input referred noise of the circuit of Fig. 3.16 as a function of the sampling capacitor $C_S$ value for different values of $NF_{LNA}$. The input signal is a rectangular pulse with $\tau$=1 ns, the amplifier voltage gain is equal to 10, the ON resistance of the switch in the S&H is equal to 50 Ω and $R_a$ is equal to 50 Ω

The minimum input referred noise voltage for a receiver using a LNA with $NF_{LNA}$=2 (3 dB) is around 32 μVrms and is obtained with a sampling capacitance value of approximately 11 pF. Increasing the voltage gain of the LNA ($G_0$) does not improve the minimum input referred noise voltage value since the input referred noise of the S&H circuit is already smaller than the thermal noise received by the antenna. It is important to notice, that the input referred noise value obtained using an LNA in the receiver (32 μVrms) is similar to the value of the input referred noise obtained using only a S&H circuit (25 μVrms for a sampling capacitance value of 10 pF). This can be explained by the fact that the S&H circuit limits the bandwidth of the noise received by the antenna, resulting that the output noise of the S&H is only dependent on the sampling capacitance value. When a LNA is used, the total S&H noise remains constant (assuming that the sampling capacitance does not change) and the noise received by the antenna (amplified by the LNA) is added to the output. The PSD of the noise from the antenna ($S_{Nin} = 4 \cdot K_B \cdot T_{amb} \cdot R_a$) is similar to the PSD from the switch resistance $R_{on}$ ($S_{N\,R_{on}} = 4 \cdot K_B \cdot T \cdot R_{on}$) because, although $R_{on}$ is smaller than $R_a$, $T_{amb}$ is smaller than $T$ (the circuit's temperature is higher than the ambient temperature), resulting that the two PSD are similar. Therefore it is expected that the input referred noise obtained when using a LNA would be similar to the input referred noise when not using a LNA. Of course, if the internal noise of the LNA is included (as it should) the input referred noise when using a LNA becomes larger than the input referred noise when not using a LNA. This is visible in the input referred noise expression (expression 3.52), if the LNA voltage gain ($G_0$) is high and the above conditions $\left(\frac{T}{T_{amb}} \approx \frac{R_a}{R_{on}}\right)$ are valid, the previous expression can be simplified to

$$V_{Nin}^2 = \frac{K_B \cdot T}{C_S} \cdot \left(0 + NF_{LNA} \times 1\right) \cdot \frac{1}{\left(1 - e^{-\tau/(R_{on} \cdot C_S)}\right)^2} \tag{3.53}$$

Comparing this simplified expression with the expression of the input referred noise of the S&H circuit, shown next (expression 3.54)

$$V_{Nin\,SH}^2 = \frac{K_B \cdot T}{C_S} \cdot \frac{1}{\left(1 - e^{-\tau/((R_{on} + R_S) \cdot C_S)}\right)^2} \tag{3.54}$$

it is clear that the input referred noise when using a LNA will always be larger than the input referred noise of the S&H circuit without the LNA, since the *NF* of the LNA has to be larger than 1 (0 dB). The main advantage of using the LNA is that the signal level at the S&H input is larger, which is useful to reduce the importance of non-ideal effects in the S&H, such as offset. The LNA can also be used to provide a controlled impedance to the antenna.

In the previous analysis, it was assumed that the LNA had a large bandwidth and small output impedance; however the design of a wide-band low noise amplifier is very challenging. Traditional LNA designs rely on tuned circuits (LC or micro-strips) to achieve impedance matching and low noise performance [17], [25]. A tuned LNA has a band-pass transfer function and the best performance is only obtained in a small frequency band around the resonating frequency of the matching circuits, resulting in a small fractional bandwidth. The LNA described in [25] has bandwidth ranging from 1 GHz to 2 GHz, resulting in a fractional bandwidth (expression 2.1) of approximately 2/3. Using matching circuits with multiple poles, built using several inductors and capacitors, it is possible to extend the fractional bandwidth of the LNA. The LNA presented in [26] uses LC ladder filters to obtain a bandwidth ranging from 3 GHz to 10 GHz, resulting in a fractional bandwidth of approximately 1.08, with the spot *NF* value ranging from 2.5 dB to 4.2 dB within the band. Another wide-band LNA design, presented in [27] uses a similar approach to obtain a bandwidth ranging from 3 GHz to 5 GHz (fractional bandwidth equal to 0.5), with the spot *NF* value ranging from 2.7 dB to 3.7 dB within the band. The frequency response of these circuits varies inside the band because it is very difficult to adjust the position of the poles in order to have a flat frequency response of the main parameters (gain, spot *NF* and input impedance). As shown in [28], even in the case of a wide band antenna, it is very difficult to design a matching circuit that guaranties a wide band impedance matching between the antenna and the LNA.

An alternative solution to extend the bandwidth of the LNA is to use distributed circuits, where several amplifiers circuits, connected by transmission lines, are used to amplify the input signal. These types of circuits have a trade off between delay and bandwidth. Since this type of amplifier uses multiple sub-amplifiers, it normally has a power and area penalty associated and a careful optimization of the number of stages must be performed during the design. The circuit described in [29] was able to achieve a bandwidth ranging from 3.1 GHz to 10.6 GHz (fractional bandwidth equal to 1.09) with a constant spot *NF* value of 2.9 dB. The frequencies response of this type of circuit is fairly flat because it uses transmission lines.

Another option is to design the LNA as a low pass-pass filter using negative feedback to adjust the input impedance and set the amplifier gain. By using a second amplifier (without feedback) and adding the two resulting signal paths together, the circuit described in [30], is capable of achieving thermal noise cancellation, resulting in a spot *NF* value below 2 dB for a bandwidth between 250 MHz and 1100 MHz (fractional bandwidth equal to 1.25).

The power consumption of the LNA is dependent on the desired bandwidth, *NF* and output impedance. Since the LNA is supposed to drive the very large capacitance of the S&H (around 10 pF), it is necessary to dissipate a large power to obtain a small value for the output impedance. The output node of the LNA would also be submitted to the charge injection and clock-feed-through from the large S&H switches, which can disturb the behavior of the circuit. In Fig. 3.19, the small signal model of a simple LNA, built using a single MOS transistor, connected to a S&H is shown.

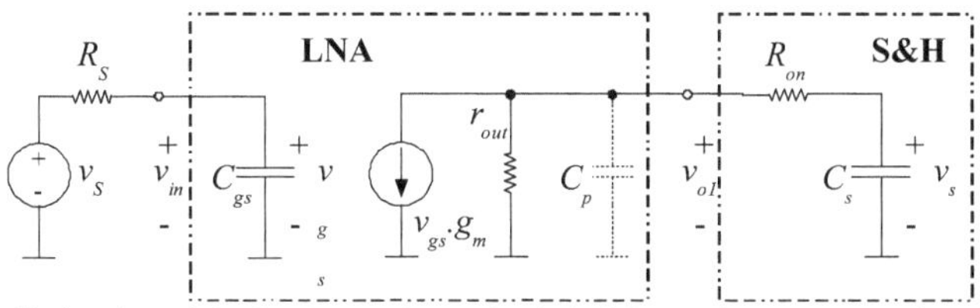

Fig. 3.19 Small signal model of a LNA using a single MOS transistor connected to the S&H

By inspecting the previous circuit, it is evident, that in order to obtain a bandwidth larger than the S&H circuit it is necessary to have a small value for $r_{out}$, since $C_p$ can be very large (it is mainly due to the large parasitic capacitance of the S&H switches). As a result, in order to obtain a reasonable voltage gain ($G_0 \approx g_m \cdot r_{out}$) it is necessary to have a large $g_m$ value, which can only be obtained by using a transistor biased with a very large drain current.

## 3.3.7 Using a Switched-Integrator

From the previous sections, two issues resulted first, the S&H circuit requires very large size transistors for the switches, in order to adequately sample the short duration input signal, the large parasitic capacitances of these transistors always cause signal attenuation (see Appendix B). The large size transistors also cause charge injections and clock feed-through. This can be a problem for averaging the input samples and for using a LNA before the S&H circuit. Second, designing a LNA with a large bandwidth and low *NF* can be very difficult and result in a large power dissipation, specially if the LNA has to drive the large sampling capacitance and the large parasitic capacitances of the S&H switches.

These two issues can be bypassed if a different approach is followed. The noise power at the output of the S&H is given by expression 3.28, which shows that making $C_s$ infinitely large would result in a noise power equal to zero. Unfortunately, this would also result in making the sampled voltage value equal to zero. However, if $C_s$ is very large the S&H transfer function can be simplified to

$$\left|\frac{1}{j \cdot 2 \cdot \pi \cdot f \cdot R_{on} \cdot C_S + 1}\right| \approx \left|\frac{1}{j \cdot 2 \cdot \pi \cdot f \cdot R_{on} \cdot C_S}\right| \tag{3.55}$$

which is the transfer function of an ideal integrator. This suggests that using an integrator transfer function instead of a low-pass transfer function can result in a lower noise at the output. In order to transform the S&H circuit into an integrator, it is necessary to convert the input signal from a voltage to a current that charges a capacitor. The input voltage can be converted to a current using a MOS transistor and its output current can be switched, during a time interval ($\tau$), into the sampling capacitor, in order to produce each output voltage sample, the clock signal has a frequency equal to $F_{SAMPLE}$. The simplified schematic of this new receiver circuit is shown in Fig. 3.20.

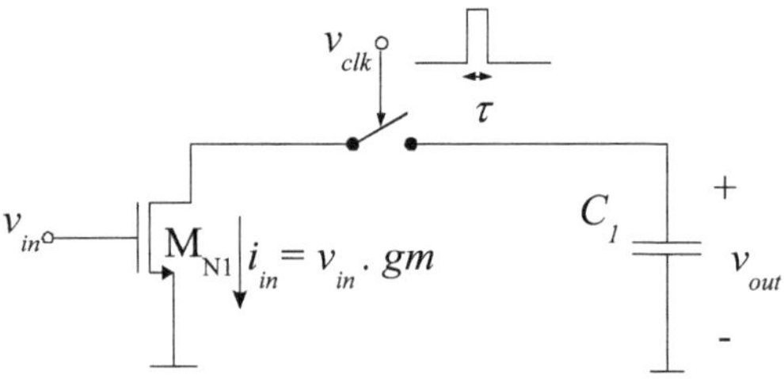


**Fig. 3.20** Simplified schematic (without bias circuits) of the switched-integrator circuit

The sampling clock signal ($v_{clk}$), controls the time instant ($t_0$) of when the switch is turned ON during the time interval $\tau$. The output voltage after the switch opens is given by

$$v_{out}(t_0) = -\int_{t_0}^{t_0+\tau} \frac{v_{in}(t)\cdot gm}{C_1} dt \tag{3.56}$$

The previous expression shows that the output voltage in the capacitor $C_1$, when the switch opens, is proportional to the average value of the input voltage during the time interval $\tau$.

Considering the case when the input signal is a rectangular pulse[8] with a duration equal to $\tau_p$ and amplitude $A_p$, the maximum output voltage at the output of the circuit is given by

$$v_{out}(t_0) = -A_p \cdot \frac{gm}{C_1} \cdot \tau_p \tag{3.57}$$

this voltage is obtained when the integration time ($\tau$) is larger or equal to the pulse duration ($\tau_p$) and the sample time is adjusted so that the time interval $\tau_p$ is inside the time interval $\tau$. From the previous expression results that the maximum output voltage can be larger than the input pulse amplitude if the ratio $gm.\tau_p/C_1$ is designed to be larger than 1. For the same input signal, the maximum output voltage of the S&H circuit (given by expression 3.33) is equal to $A_p$. Therefore using an integrator, instead of a S&H circuit, has the advantage of providing gain in the signal path. However, the integrator changes the shape of the received signal, but since the transmitted pulses are differentiated when they are radiated by the transmission antenna (see Section 2.3), the integration performed by the receiver cancels this differentiation. Integrating the received signal can also reduce the time dispersion, as shown in Fig. 2.12 (first derivative of a Gaussian pulse) and

[8] As previously mentioned, this type of pulse facilitates the subsequent mathematical analysis of the circuit.

Fig. 2.13 (second derivative of a Gaussian pulse) of the previous chapter. The second derivative of the Gaussian pulse has a shorter duration than the third derivative of the Gaussian pulse and if the received pulse has a shorter duration this will improve the accuracy of the ranging operation.

In order to obtain a more exact expression for the maximum output voltage of the circuit, it is necessary to include the impedance of the input voltage source and the parasitic $C_{gs}$ capacitance of the transistor $M_1$, as shown in Fig. 3.21.

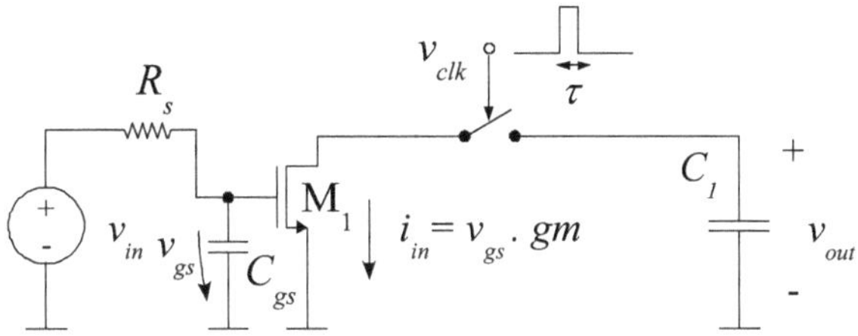


**Fig. 3.21** Simplified schematic of the receiver circuit including the input source resistance

Analyzing the circuit, when the switch is closed, results in the following *s* transfer function

$$\frac{V_{out}(s)}{V_{in}(s)} = -\frac{gm}{s \cdot C_1} \cdot \frac{1}{1 + s \cdot R_S \cdot C_{gs}} \tag{3.58}$$

The output voltage in the time domain, can be calculated by applying the inverse *Laplace* transformation to the following expression

$$V_{out}(s) = V_{in}(s) \cdot \left( \frac{V_{out}(s)}{V_{in}(s)} \right) \tag{3.59}$$

where $V_{in}(s)$ is the *Laplace* transform of the input signal, in this case a rectangular pulse with a duration equal to $\tau_p$ and amplitude $A_p$. The rectangular pulse can be mathematically defined as the sum of two step functions, where the second step function has a negative amplitude ($-A_p$) and a time delay equal to the pulse duration ($\tau_p$). The output voltage will be constituted by the sum of the circuit's response to each of the step functions. The step response can be calculated by applying the inverse *Laplace* transform to expression 3.59 (for a step input) which results in the following expression

$$v_{out}(t) = -\frac{A_p \cdot gm}{C_1} \cdot \left( t + C_{gs} \cdot R_S \left( e^{\frac{-t}{C_{gs} \cdot R_S}} - 1 \right) \right) \tag{3.60}$$

Two output voltages of the circuit, for different values of $C_{gs}$, are plotted in Fig. 3.22, for a rectangular pulse with $A_p$=1 V and $\tau_p$=1 ns. The value of $gm$ is 6 mS, the value of $C_1$ is 5 pF and the value of $R_s$ is 50 Ω.

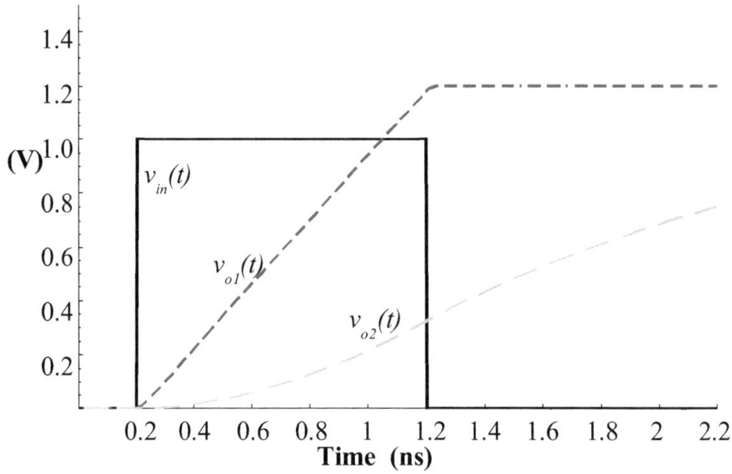


**Fig.** 3.22 Output voltages of the integrator for a rectangular pulse input voltage ($v_{in}(t)$) with $A_p$=1 V and $\tau_p$=1 ns. The $v_{o1}(t)$ output voltage is for a circuit with $C_{gs}$=0.3 pF and the $v_{o2}(t)$ output voltage is for a circuit with $C_{gs}$=30 pF

The previous graph shows that if the time constant $R_s.C_{gs}$ is small when compared to the pulse duration ($\tau_p$) (curve $v_{o1}(t)$), the output voltage ($v_{o1}(t)$) is approximately equal to the ideal integrator output voltage (expression 3.57). The maximum output voltage can be obtained if the switch is closed at (or before) the beginning of the input pulse and opened at (or after) the end of the input pulse (corresponding to $\tau = \tau_p$). This example also shows, that it is possible to obtain a larger output amplitude than the amplitude of the input pulse, by choosing a correct set of values for the circuit's components. The maximum voltage gain of this circuit, when the switch is closed during a time interval $\tau$ coinciding with the input pulse duration, can be defined as

$$G = \frac{v_{out}(\tau)}{A_p} = -\frac{gm}{C_1} \cdot \left( \tau + C_{gs} \cdot R_S \left( e^{\frac{-\tau}{C_{gs} \cdot R_S}} - 1 \right) \right) \quad (3.61)$$

The output voltage ($v_{o1}(t)$) shown in the previous figure is from an integrator with the switch closed. In order to produce an output sample it is necessary to close and open the switch at appropriate time instants. First, the output capacitor is reseted and the switch is closed, after a time interval $\tau$ the switch is opened and an output sample is obtained. As previously discussed, the input signal is reconstructed by producing samples controlled by a sample clock with a progressively increasing delay (see Section 3.3.3). There are two options to consider for the duration of time that the switch remains closed ($\tau$) $\tau >> \tau_p$ or $\tau \leq \tau_p$. This is illustrated in Fig. 3.23, where the two cases are depicted. This figure depicts the output voltage samples that are obtained for the two types of clock signals that control the integrator switch. The switch is closed when the clock is high and opens on the falling edge of the clock, producing the output sample. The output samples, produced by each clock, are obtained after the integrating capacitor is reseted and are obtained for independent sweeps of the input signal. Note that both cases produce the same $v_3$ sample, but in the case when $\tau >> \tau_p$ the samples produced after $v_3$ all have the same value. This is because of the memory effect of the integrator and it means that if more than one signal is present during the time interval $\tau$, the output samples would have a contribution from all these signals. This could difficult the analysis of distant echoes since the first echoes to arrive generally have a larger amplitude than the last echoes to arrive. By using $\tau \leq \tau_p$ it is possible to eliminate this memory effect and produce samples that are only depend on the current echo.

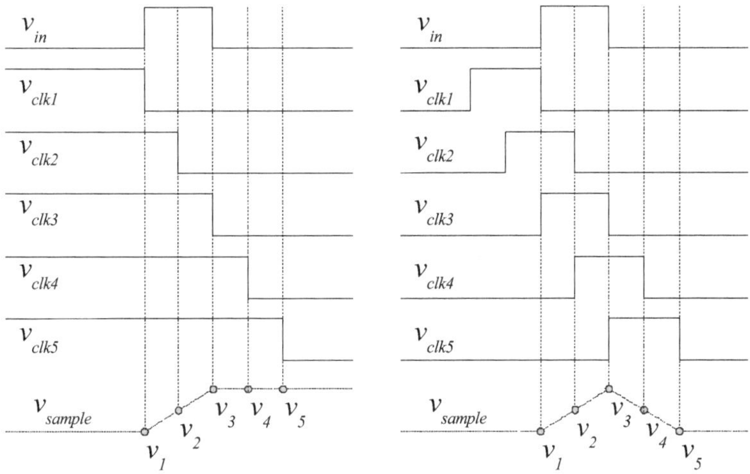


**Fig.** 3.23 Switched integrator output samples for a rectangular pulse input signal obtained with 5 clock signals with increasing delay. On the left side the integrator switch is turned ON for a long time prior to the falling edge of the clock. On the right side the switch is turned ON during a time interval equal to the pulse duration

In the left side of the figure, the clock signal is high for a long time before the falling edge and on the right side of the figure; the clock is high only for a time interval equal to the input pulse duration. The different output samples ($v_1$ to $v_5$) are obtained by using increasing delays in the clock signal.

There are several noise sources in the circuit of Fig. 3.21 the thermal noise of the source resistance ($R_s$), the thermal noise of the transistor $M_1$[9] channel and the thermal noise of the ON resistance of the switch ($R_{ON}$). Considering that the value

[9] For frequencies below 1 GHz, the induced gate noise is much smaller than the channel thermal noise and it can be ignored in the noise analysis.[31].

of the $r_{ds}$ resistance of transistor $M_1$ is much larger than the value of $R_{ON}$, the contribution of the latter resistance to the output noise power is negligible.

The channel noise of transistor $M_1$ is due to the channel resistance and it can be represented by a current source in parallel with the drain and source nodes of the transistor. This current source has a PSD equal to $\gamma \cdot 4 \cdot K_B \cdot T_{amb} \cdot g_{d0}$, where $g_{d0}$ is the drain-source conductance for a $V_{DS}$ voltage equal to 0 V ($g_{d0} \approx gm$) and $\gamma$ is a factor that models the excess noise due to the large electrical field that appear in short channel transistors, it depends on the transistor channel length and the value of the $V_{DS}$ voltage.[10] For long channel devices, $\gamma$ is equal to 2/3 and for short channel transistors it can be as large as 1.8 [31], [32].

In order to calculate the contribution of the transistor channel noise to the output noise power it is necessary to calculate the contribution of the transistor noise current source into the output voltage. This is done analyzing the following circuit (Fig. 3.24).

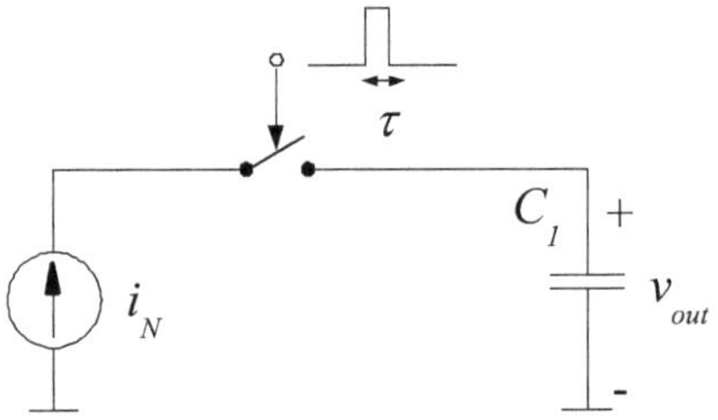


**Fig. 3.24** Circuit used to calculate the channel noise contribution into the total output noise

The previous circuit is a linear time-varying system, because its input ($i_N$) to output ($v_{out}$) transfer function is not invariant when a time translation is applied to the input signal. Since the switch is closed during the time interval $\tau$ and open outside this interval, it is clear that the circuit's impulse response is different depending on the excitation time instant ($t_0$). Therefore, the impulse response of the circuit ($h(t,t_0)$ ) is a function of both the observation time ($t$) and the excitation time instant ($t_0$) [33]. The impulse response of the circuit is equal to the derivative of the circuit step response [18]. The step response of the circuit, when the switch is closed, can be obtained from expression 3.60 by making $A_p \cdot gm = I_N$ and $C_{gs} = R_S = 0$ resulting in

$$v_{out}(t) = \frac{I_N}{C_1} \cdot t \tag{3.62}$$

The step response of the circuit, with the switch closing at $t=-\tau/2$ and opening at $t=\tau/2$, for a step input occurring at $t=t_0$ is given by

[10] The value of the $V_{DS}$ voltage should be kept as low as possible in order to minimize the electrical field inside the transistor channel and thus reduce the transistor noise factor.

$$v_{out}(t,t_0) = \begin{cases} 0 & if\ t < t_0 \\ 0 & if\ t < -\tau/2 \\ \frac{I_N}{C_1}\cdot(t-t_0) & if\ t_0 < t < -\tau/2 \\ \frac{I_N}{C_1}\cdot(\tau-t_0) & if\ t > \tau/2 \\ 0 & if\ t_0 > \tau/2 \end{cases} \tag{3.63}$$

From the previous expression results that the impulse response of the circuit is given by

$$h(t,t_0) = \frac{\frac{d}{dx}v_{out}(t,t_0)}{I_N} = \begin{cases} 0 & if\ t < t_0 \\ 0 & if\ t < -\tau/2 \\ \frac{1}{C_1} & if\ t_0 < t < -\tau/2 \\ 0 & if\ t > \tau/2 \\ 0 & if\ t_0 > \tau/2 \end{cases} \tag{3.64}$$

The transfer function can be obtained by applying the Fourier transform to expression 3.64, resulting in

$$H(f,t_0) = \int_{-\infty}^{+\infty} h(t,t_0)\cdot e^{-j\cdot 2\cdot\pi\cdot t\cdot f}dt = \begin{cases} 0 & if\ t_0 < \frac{-\tau}{2} \\ \frac{1}{C_1}\cdot\frac{e^{-j\cdot 2\cdot\pi\cdot f\cdot\tau/2} - e^{-j\cdot 2\cdot\pi\cdot t_0\cdot f}}{-j\cdot 2\cdot\pi\cdot f} & if\ \frac{-\tau}{2} < t_0 < \frac{\tau}{2} \\ 0 & if\ t_0 > \frac{\tau}{2} \end{cases} \tag{3.65}$$

Noting that the current source $I_N$ is connected when the switch closes, the excitation time $t_0$ is equal to $-\tau/2$ and the transfer function is given by

$$H\left(f,\frac{-\tau}{2}\right) = \frac{1}{C_1}\cdot\frac{e^{-j\cdot 2\cdot\pi\cdot f\cdot\tau/2} - e^{j\cdot 2\cdot\pi\cdot f\cdot\tau/2}}{-j\cdot 2\cdot\pi\cdot f} = \frac{1}{C_1}\cdot\frac{\sin(\pi\cdot f\cdot\tau)}{\pi\cdot f} \tag{3.66}$$

This transfer function is a *sinc* function and it is equal to zero at frequencies multiple of $(1/\tau)$. The noise power at the present at the output voltage due to the transistor channel noise is calculated using

$$V_{NM1}^2 = \int_0^{+\infty} \gamma \cdot 4 \cdot K_B \cdot T \cdot gm \cdot \left| H\left(f, \frac{-\tau}{2}\right) \right|^2 df = \frac{\gamma \cdot 2 \cdot K_B \cdot T \cdot gm \cdot \tau}{C_1^2} \quad (3.67)$$

This noise power can be referred to the input dividing expression 3.67 by the circuit gain squared (expression 3.61 squared), resulting in the following expression for the input referred noise

$$V_{Nin\,M1}^2 = \frac{\gamma \cdot 2 \cdot K_B \cdot T \cdot \tau}{gm \cdot \left( \tau + C_{gs} \cdot R_S \cdot \left( e^{-\tau / C_{gs} \cdot R_S} - 1 \right) \right)^2} \quad (3.68)$$

The previous expression shows that the input referred noise voltage does not depend on the value of the $C_1$ capacitance. If the time constant, associated with input low-pass filter formed by $R_s$ and $C_{gs}$, becomes larger than $\tau$, the input referred noise will increase, since the gain (given by expression 3.61) decreases. If the time constant ($R_s.C_{gs}$) is much smaller than $\tau$, the previous expression can be simplified to

$$V_{Nin\,M1}^2 \approx \frac{\gamma \cdot 2 \cdot K_B \cdot T}{gm \cdot \tau} \quad (3.69)$$

This shows that the input referred noise voltage can be reduced by increasing the value of the $gm$ of the transistor $M_1$ or by increasing the time interval ($\tau$) when the switch is closed. However it is important to remember that the output noise power grows linearly with $\tau$, but the gain only grows linearly with $\tau$ until $\tau = \tau_p$, remaining constant after that (see Fig. 3.22). If $\tau$ is increased beyond that value the input referred noise will start to increase.

The contribution of the thermal noise of the source resistance ($R_s$) to the output noise power can be calculated in a similar manner. The low pass circuit formed by $R_s$ and $C_{gs}$ (see Fig. 3.23), filters the PSD of the $R_s$ thermal noise, resulting in a noise voltage across the gate of the transistor with a PSD equal to

$$V_{gs\,N\,R_S}^2 \cdot \Delta f = 4 \cdot K_B \cdot T \cdot R_S \cdot \left| \frac{1}{1 + C_{gs} \cdot R_S \cdot j \cdot 2 \cdot \pi \cdot f} \right|^2 \quad (3.70)$$

The voltage $V_{gs}$ is converted into a current by the transistor transconductance and this current is applied to the $C_1$ capacitance when the switch is closed. Therefore, the noise power present at the output, due to the resistor $R_s$ can be calculated using

$$V_{N R_S}^2 = \int_0^{+\infty} 4 \cdot K_B \cdot T \cdot R_S \cdot \left| \frac{gm}{1 + C_{gs} \cdot R_S \cdot j \cdot 2 \cdot \pi \cdot f} \right|^2 \cdot \left| \frac{1}{C_1} \cdot \frac{\sin(\pi \cdot f \cdot \tau)}{\pi \cdot f} \right|^2 df = \\ = \frac{2 \cdot K_B \cdot T \cdot gm^2 \cdot R_S}{C_1^2} \cdot \left( \tau + C_{gs} \cdot R_S \cdot \left( e^{-\tau / C_{gs} \cdot R_S} - 1 \right) \right) \tag{3.71}$$

This noise power can be referred to the input by dividing expression 3.71 by the circuit gain squared (expression 3.61 squared), resulting in the following expression for the input referred noise

$$V_{Nin R_S}^2 = \frac{2 \cdot K_B \cdot T \cdot R_S}{\left( \tau + C_{gs} \cdot R_S \cdot \left( e^{-\tau / C_{gs} \cdot R_S} - 1 \right) \right)} \tag{3.72}$$

Once again, the input referred noise does not depend on the value of $C_1$ and if the time constant, associated with input low-pass filter formed by $R_s$ and $C_{gs}$ increases, the gain decreases and thus the input referred noise increases. If this time constant becomes much smaller than $\tau$, the previous expression can be simplified to

$$V_{N R_S}^2 \approx \frac{2 \cdot K_B \cdot T \cdot R_S}{\tau} \tag{3.73}$$

This expression shows that the input referred noise due to the source resistance can be reduced by increasing the integration time $\tau$. The total input referred noise of the circuit is obtained by adding expressions 3.68 and 3.72, resulting in

$$V_{N in}^2 = V_{N in\, M1}^2 + V_{N in\, R_S}^2 \tag{3.74}$$

The total input referred noise value does not depend on the value of the integrating capacity $C_1$ but it depends on the $\tau$, $C_{gs}$ and $gm$ values. The last two parameters ($C_{gs}$ and $gm$) depend on the transistor size and transistor bias current ($I_{DM1}$). The transistor transconductance ($gm$) can be calculated using

$$gm(I_D, V_{Dsat}) = \frac{2 \cdot I_D}{V_{Dsat}} \tag{3.75}$$

The transistor transconductance can be maximized, for a given bias current value, by decreasing the value of $V_{Dsat}$. The transistor gate capacitance ($C_{gs}$) can be calculated using

$$C_{gs}\left(I_D, V_{Dsat}\right) = \frac{2}{3} \cdot \frac{C_{ox} \cdot 2 \cdot I_D \cdot L^2}{K_N \cdot V_{Dsat}^2} \tag{3.76}$$

where $C_{ox}$ and $K_N$ are technology dependent parameters. This expression shows, that in order to minimize the capacitance $C_{gs}$ ,it is very important to use the smallest transistor channel length ($L$) available in the technology, since this capacitance increases with the square of $L$. Selecting a larger $V_{Dsat}$ value also results in a smaller capacitance value for a given bias current.

Substituting $C_{gs}$ and $gm$ in expressions 3.68 and 3.72 by expressions 3.75 and 3.76, results in an expression for the input referred noise which is a function of the current $I_{DM1}$ and $V_{Dsat}$ voltage of the transistor $M_1$. The channel length ($L$) is set to the minimum value allowed by the technology in order to reduce the $C_{gs}$ capacitance to the minimum value possible. In Fig. 3.25, the input referred noise value is plotted for $I_{DM1}$ changing from 0.1 mA to 100 mA and three different $V_{Dsat}$ values (100 mV, 200 mV and 300 mV). Again, the temperature value is equal to 290 Kelvin and the $R_s$ value is equal to 50 Ω.

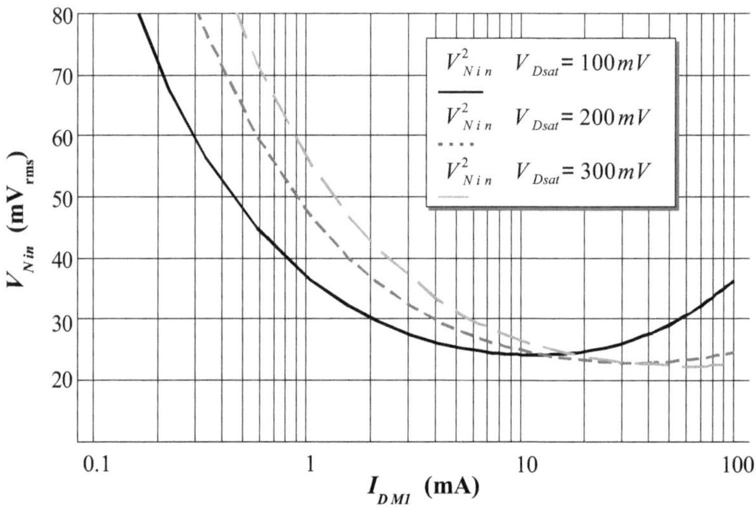


**Fig. 3.25** Input referred noise of the switched integrator circuit for different values of $I_D$ and $V_{Dsat}$. The input signal is a rectangular pulse with $\tau$=1 ns, the $R_S$ value is 50 Ω

The previous graph shows that the input referred noise of the switched integrator reduces with increasing values of the transistor bias current, this is because increasing the bias current increases the value of the transconductance of the transistor. For a given value of bias current, the input referred noise is reduced for smaller values of $V_{Dsat}$ because the $gm$ value increases when $V_{dsat}$ decreases. However, for larger values of the bias current the input referred noise starts to increase, this occurs because the transistor parasitic capacitance becomes very large and therefore the gain starts to decrease, thus reducing the input referred noise value. As expected, if the $V_{Dsat}$ voltage is increased, this effect will occur for larger bias currents because the $C_{gs}$ capacitance is smaller for a given value of bias current when the $V_{Dsat}$ voltage increases. The graph also shows that the minimum value of the input referred noise (around 22 μV) is obtained by using a large value for the $V_{Dsat}$ voltage (>300 mV) and using a large value for the bias current (>100 mA). However if a smaller $V_{Dsat}$ voltage is selected (100 mV), the minimum input referred noise value is 25 μV for a bias current value of approximately 10 mA. An improvement of 3 μV in the input referred noise costs an increase of 10 times in the bias current, therefore it is better to use a smaller $V_{Dsat}$ voltage and bias current. The minimum input referred noise (25 μV) obtained using the switched-integrator circuit is similar to the minimum input referred noise obtained using the ideal S&H circuit (see Section 3.3.4). In reality, the input referred noise of the switched-integrator circuit is smaller than the input referred noise of a real S&H circuit with MOS switches because of the signal attenuation introduced by large parasitic capacitances of the MOS switches. The switched-integrator circuit, has also the advantage of providing a gain larger than 1 in the signal path, when compared with the S&H circuit. The switched-integrator can also be used to add samples together, in order to improve the input referred noise, as discussed in Section 3.3.5. If for example 100 samples are added together the input referred noise would be reduced from 25 μV to 2.5 μV.

### 3.3.8 Differential Switched-Integrator Circuit

The ON resistance value of the switch used in the switched-integrator is not critical, since this switch works with a current signal instead of voltage signal. This means that the MOS transistors that implement this switch do not need to be very large, as it was the case with the transistors used to implement the switch in the S&H circuit (Appendix B). However, when the switch opens, it is necessary to provide an alternative path for the $I_{DM1}$ current to flow. This is done using a second switch with a complementary control signal that dumps the input current to $V_{DD}$. This type of arrangement is known as a current steering switch and it is basically a differential pair. This is a very fast and efficient way of switching current signals. This is the approach that is followed in the design of very fast digital to analog converters (DAC), the fastest DAC architectures reported in the literature [34], [35] are based on current steering circuits. A simplified version of the schematic of the switched-integrator circuit using a current steering switch is shown next in Fig. 3.26.

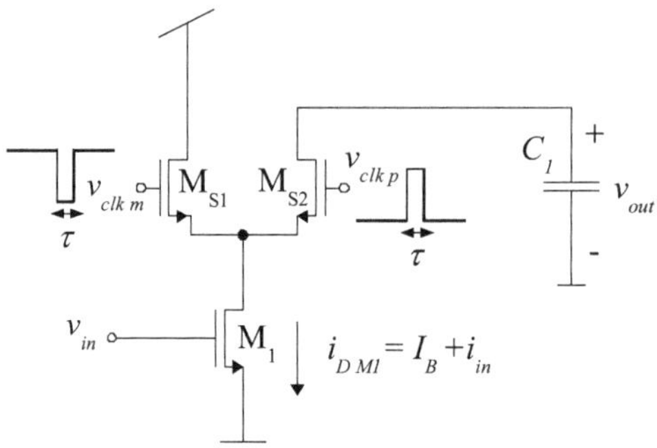


**Fig. 3.26** Simplified schematic of the switched-integrator circuit using a current steering switch

The differential pair works essentially in saturation, where the current $i_{DM1}$ either flows through transistor $M_{S1}$ (the switch is OFF) or $M_{S2}$ (the switch is ON), depending on the clock signal. The clock signal is a differential signal with voltage levels that guaranty that when the transistor $M_{S1}$ is ON, the transistor $M_{S2}$ is OFF (or vice versa). The circuit that generates the differential clock signal has to produce a short pulse with a time duration equal to $\tau$. This circuit will be addressed in detail in the next chapters.

The thermal noise from the channel of transistor $M_{S2}$ does not influence significantly the output noise because this device is a cascode device and therefore the current from its noise source is attenuated by $\left( \approx {gds_{MN1}}\big/{gm_{MS2} + gds_{MS2} + gds_{MN1}} \right)$, because of the cascode configuration.

The parasitic capacitances ($C_{sb}$ and $C_{db}$) of the transistor $M_{S1}$ and $M_{S2}$, do not introduce significant signal attenuation, although they are in the signal path. The $C_{sb}$ capacitances of transistors $M_{S1}$ and $M_{S2}$ are connected to the source node of the differential pair which has a very small voltage swing due to the cascode effect of transistor $M_{S2}$. The $C_{db}$ capacitance of transistor $M_{S2}$ is in parallel with the output capacitor and therefore its attenuation can be eliminated by making $C_1$ smaller.

The circuit shown in Fig. 3.26 has a fundamental problem that prevents its intended use. The output capacitor ($C_1$) integrates the current $i_{D\,M1}$ , therefore it will also integrate the bias current ($I_B$). Since the bias current is very high (typically larger than 1 mA as shown in the graph of Fig. 3.25) and the signal current ($i_n$) is expected to be very small, the resulting output voltage will be essentially a large DC offset due to the bias current. In order to solve this problem it is necessary to use a fully differential structure for the switched-integrator, this structure is shown in Fig. 3.27.

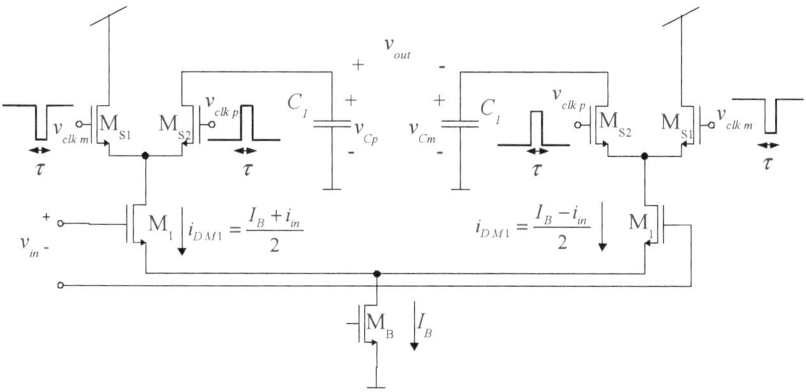


**Fig. 3.27** Differential circuit of the switched-integrator circuit (Gilbert multiplier)

The $M_1$ transistor is now part of a differential pair with a bias current equal to $I_B$. The input signal is applied to this differential pair, producing a differential current that it is switched by two differential pairs between $V_{DD}$ and the two capacitors ($C_1$). This circuit is known as the Gilbert multiplier [36] and normally is used with sinusoidal signals, both for the clock signal and the input signal. The typical use of this circuit is as mixer to down-convert or up-convert base-band signals to RF or IF frequencies. In this case, the clock signals $v_{clkp}$ and $v_{clkm}$ are a differential clock signal ($v_{clk}$) that are created by a differential pulse generator, that will be described in the next chapters.
The voltage in each of the $C_1$ capacitors ($v_{Cp}$ and $v_{Cm}$) when the switch opens is equal to (assuming that $i_{in} << I_B$ and that $R_s.C_{gs} << \tau$)

$$v_{C_m^p}(\tau) = \frac{-I_B \cdot \tau}{2 \cdot C_1} \pm \frac{i_{in} \cdot \tau}{2 \cdot C_1} \tag{3.77}$$

The common mode output voltage variation is equal to

$$v_{o\,CM}(\tau) = \frac{v_{Cp}(\tau) + v_{Cm}(\tau)}{2} = \frac{-I_B \cdot \tau}{2 \cdot C_1} \tag{3.78}$$

and the differential output voltage variation is equal to

$$v_o(\tau) = v_{Cp}(\tau) - v_{Cm}(\tau) = \frac{i_{in} \cdot \tau}{C_1} \tag{3.79}$$

The two previous expressions show that the differential output voltage only depends of the input signal and that the large bias current is translated into a common mode negative voltage step. The common mode output voltage value must be large enough to guaranty that transistors $M_B$, $M_1$ and $M_{S2}$ are in the saturation region ($V_{o\,CM} > V_{DS\,B} + V_{DS\,M1} + V_{DS\,S2}$). If the common mode voltage step is too large, this condition could be violated and the circuit would not function properly. Therefore the value of the $C_1$ capacitance must be sufficiently large to make the common mode voltage drop small (expression 3.78). This is especially important in the case of a low power supply voltage, where the available voltage margin in the $V_{DS}$ voltages of the transistors is small. If a maximum voltage drop value equal to $\Delta V_{CMmax}$ is tolerated, the minimum allowable capacitance value of $C_1$ is given by

$$C_1 \geq \frac{I_B \cdot \tau}{2 \cdot \Delta V_{CM\,\max}} \tag{3.80}$$

The gain of the switched-integrator is given by expression 3.61, considering that the condition $R_s.C_{gs} << \tau$ is applicable, the expression for the maximum possible gain of the circuit is obtained replacing $C_1$ by expression 3.80 and *gm* by expression 3.75, resulting in

$$G_{\max} = \frac{4 \times \Delta V_{CM\,\max}}{V_{Dsat}} \tag{3.81}$$

Considering that $V_{Dsat}$=100 mV and $\Delta V_{CMax}$=200 mV, results that the maximum gain value of the switched-integrator would be equal to 8.

The previous analysis shows a critical problem with the circuit depicted in 3.27, each time the differential switch closes the output common voltage will drop by $\Delta V_{CMmax}$, so after a given number of clock pulses, the common mode output voltage will be zero and the circuit will stop working. In order to solve this problem is necessary to add an extra circuit that can supply a positive current into the output capacitors, so that the output common mode voltage value is, on average, constant. This circuit should only adjust the common mode voltage and leave the differential voltage intact. This circuit is constituted by two transistors working as current sources, both with a bias current value ($I_{B2}$), such that, during the period of the sample clock, it will charge the output capacitors with a voltage equal to the common mode voltage drop that occurs during the sampling ($\Delta V_{CMax}$). The value of $I_{B2}$ must be adjusted by a feedback loop that compares the common mode output voltage with the desired common mode reference voltage. This circuit is shown in Fig. 3.28.

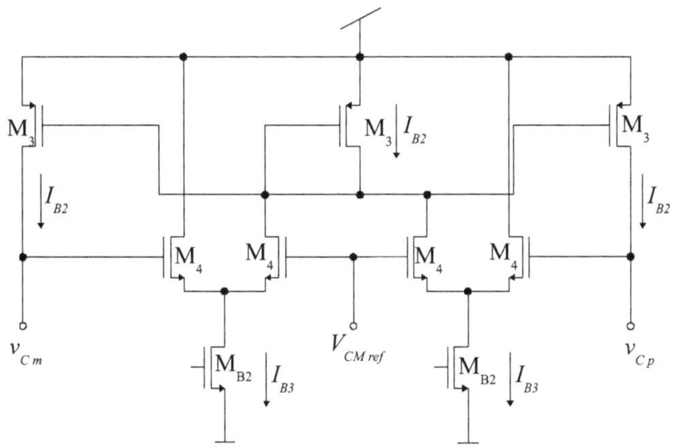


**Fig. 3.28** Switched-integrator load and common mode feedback circuit schematic

The transistors $M_3$ are the loads of the switched-integrator circuit and supply the required bias current ($I_{B2}$) necessary to maintain the common mode output voltage around the desired value ($V_{CM\,ref}$). This is achieved by using a feedback loop comprised of two differential pairs (constituted by the transistors $M_4$) that compare each single-ended output voltage ($v_{Cp}$ and $v_{Cm}$) with the $V_{CM\,ref}$ voltage. The current produced by these two differential pairs $I_{B2}$, is mirrored to the load transistors. The average value of $I_{B2}$ is

$$I_{B2} \approx \frac{I_B}{2} \cdot \frac{\tau}{T_{SAMPLE}} \tag{3.82}$$

For $\tau$=1 ns and $T_{SAMPLE}$=1 μs results that the average value of this current is 1000 times smaller than $I_B$. The bias current of the common mode feedback circuit ($I_{B3}$) is selected to be between 50 μA and 200 μA, in order to provide enough current drive capability to the circuit, immediately after the switch closes.

Considering only common mode signals, this circuit has 3 poles the first is associated to the parasitic capacitance in the source node of the differential pairs $\left(p_1 \approx gm_4/\left(2 \cdot C_{gs4}\right)\right)$, the second is associated with the parasitic capacitance in the gate node of the transistors $M_3$ $\left(p_2 \approx gm_3/\left(3 \cdot C_{gs3}\right)\right)$ and the third is associated with the capacitance of the output nodes ($v_{Cp}$ and $v_{Cm}$) $\left(p_3 \approx gds_3/C_1\right)$. The last one is clearly the dominant pole because of the large capacitance value of $C_1$ and the small conductance value of the *gds* of the transistors $M_3$. The circuit is stable because the DC gain is small $\left(\approx gm_4/gds_3\right)$ and the poles have a large separation in the frequency domain. The pole $p_2$ must have a frequency at least 4 times higher than the sample frequency, so that the circuit is able to adjust the value of $I_{B2}$ fast enough to compensate the voltage drop that occurs when the switched integrator is turned ON. The design criteria for this circuit is as follows

- $I_{B3}$ This current must be large enough to guaranty that $I_{B2}$ can be as large as the value given by expression 3.82, therefore $I_{B3} > I_B \cdot \tau / T_{SAMPLE}$
- $M_4$ designed to have a large *gm* in order to increase the gain of the feedback loop. The channel length of this transistor is equal to 0.18 μm and the $V_{Dsat}$ voltage is smaller than 130 mV.
- $M_3$ designed to have a compromise between a small *gds*, a small *gm* (to reduce the noise) and a small $C_{gs}$ to increase the $p_2$ pole frequency. The channel length is 1 μm and the $V_{Dsat}$ voltage is smaller than 300 mV.

The noise generated by the load circuit must be included in the total output noise of the switched-integrator. All the transistors in the schematic of Fig. 3.28 have thermal noise in their channels that contributes to the noise voltage present at each output node ($v_{Cp}$ and $v_{Cm}$). However, since the output sample is a differential signal i.e. it is the difference between these two voltages, any common mode noise will be canceled. Therefore, the only transistors from this circuit that contribute to the differential output noise are the load transistors $M_3$. The noise from the diode connected transistor $M_3$ appears as common mode noise. The output differential noise power generated by the two $M_3$ load transistors is given by the following expression[11]

$$V_{NM3}^2 = 2 \times \int_0^{+\infty} \gamma \cdot 4 \cdot K_B \cdot T \cdot gm_3 \cdot \left| \frac{1}{gds_3 + C_1 \cdot j \cdot 2 \cdot \pi \cdot f} \right|^2 df = \frac{\gamma \cdot 2 \cdot K_B \cdot T}{C_1^2} \cdot \frac{gm_3}{gds_3} \tag{3.83}$$

This expression shows that the noise power created by the transistors $M_3$ can be very large, since the *gm*/*gds* ratio in a transistor normally varies between 20 and 100, depending on the design ($V_{Dsat}$ and $L$) and it does not change much with the transistor bias current ($I_{B2}$). This noise power can be (much) larger than the noise power due to transistor $M_1$, which has a much larger bias current value. This is because the transistors $M_3$ are permanently connected to the output node, while the transistor $M_1$ is only connected during the short time interval $\tau$. The contribution of these transistors to the output noise power, can be reduced if the load transistors are connected to the output only during a time interval $\tau_2$ such that $\tau < \tau_2 < T_{SAMPLE}$. The impulsive response from the noise current source of the transistors to the output voltage is given by

[11] This expression, assumes that the impedance of the output node is only due to $M_3$ because transistor $M_{S2}$ is OFF.

$$h_3(t,t_0)=\begin{cases} 0 & \text{if } t<t_0 \\ 0 & \text{if } t<-\tau/2 \\ \dfrac{e^{-\frac{gds_3}{C_1}(t-t_0)}}{C_1} & \text{if } t_0<t<-\tau/2 \\ 0 & \text{if } t>\tau/2 \\ 0 & \text{if } t_0>\tau/2 \end{cases} \tag{3.84}$$

Assuming that $gds_3/C_1 << \tau_2$ is valid, then $e^{-gds_3/C_1 \cdot \tau_2} \approx 1$ holds and the impulsive response can be approximated by expression 3.64 and the same analysis that was performed to calculate the output noise power due to $M_1$ can be applied to obtain the output noise power due to $M_3$, resulting in the following expression for the differential output noise power due to the two $M_3$ transistors

$$V_{NM3}^2 = 2\times \int_0^{+\infty} \gamma \cdot 4 \cdot K_B \cdot T \cdot gm_3 \cdot \left|H_3\left(f,-\tau_2/2\right)\right|^2 df = \frac{\gamma \cdot 4 \cdot K_B \cdot T \cdot gm_3 \cdot \tau_2}{C_1^2} \tag{3.85}$$

The differential output noise power due to the two $M_1$ transistors of the differential pair can be obtained multiplying expression 3.67 by two, resulting in

$$V_{NM1}^2 = 2\times \int_0^{+\infty} \gamma \cdot 4 \cdot K_B \cdot T \cdot gm_1 \cdot \left|H_1\left(f,-\tau/2\right)\right|^2 df = \frac{\gamma \cdot 4 \cdot K_B \cdot T \cdot gm_1 \cdot \tau}{C_1^2} \tag{3.86}$$

Since the drain current of $M_3$ is (on average) equal to $I_{DM3} \approx I_{DM3} \cdot \dfrac{\tau}{\tau_2}$ (from expression 3.82), results that

$$V_{N\,M3}^2 = V_{N\,M1}^2 \cdot \frac{V_{Dsat\,M1}}{V_{Dsat\,M3}} \tag{3.87}$$

The previous expression shows that the $V_{Dsat}$ voltage of transistor $M_3$ should be as large as possible and that the $V_{Dsat}$ voltage of transistor $M_1$ should be as small as possible in order to minimize the output noise due to transistor $M_3$.

In order to disconnect the load transistors from the output nodes ($v_{Cp}$ and $v_{Cm}$), it is necessary to introduce two new transistors ($M_{S3}$), that work as switches, between the load circuit and the output capacitors, as shown in Fig. 3.29. This figure also shows the clock signals that control both the multiplier and load switches ($M_{S3}$). The $v_{clk}$ signal is a differential signal and the $v_{clk2}$ is an active low CMOS logic signal.

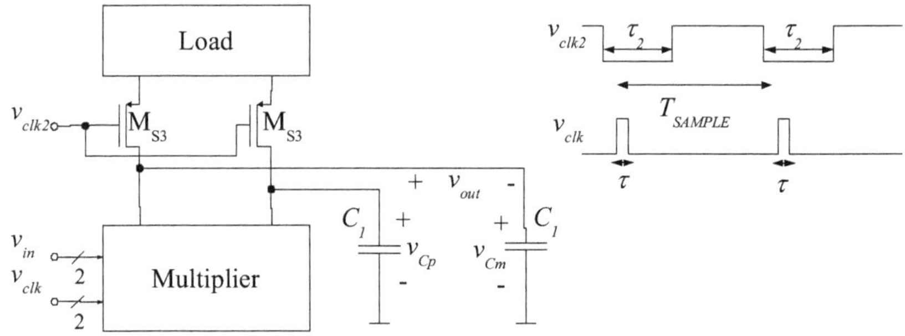


**Fig. 3.29** Switched-integrator diagram, constituted by the multiplier and load circuits together with the load switches

The load switches are designed to work in the linear region, with a small ON resistance value. The contribution of the transistors $M_{S3}$ to the output noise power is given by

$$V_{N\,MS}^2 = 2\times\int_0^{+\infty}\gamma\cdot 4\cdot K_B\cdot T\cdot R_{on}\cdot\left|\frac{1}{1+C_1\cdot\left(R_{on}+rds_3\right)\cdot j\cdot 2\cdot\pi\cdot f}\right|^2 df = \\ =\frac{2\cdot K_B\cdot T}{C_1}\cdot\frac{R_{on}}{R_{on}+rds_3} \tag{3.88}$$

which very can be small, when compared with the contribution of the other devices, if the transistors $M_S$ are designed to have an ON resistance value ($R_{ON}$) much smaller than the drain resistance of the transistors $M_3$ ($rds_3$). Therefore the contribution of the noise power from these transistors into the circuit's total noise power can be considered negligible.

### 3.3.9 Offset Calibration

As previously described in Section 3.3.5, the output samples produced by the switched-integrator can be added together, in order to reduce the input referred noise. This is achieved, simply by repeatedly sampling the input signal with the same delay without reseting the output capacitors $C_1$. In each sampling instant the switched-integrator produces a differential voltage variation, with an amplitude given by expression 3.56, that is added to the previous output voltage. But, in each sampling cycle, the switched-integrator also adds a noise signal to the output voltage. The end result, as explained in Section 3.3.5, is that after adding $n$ samples together, the signal power grows with $n^2$ and the noise power only grows with $n$, resulting that the input referred noise reduces with $n$. However, the offset voltage present in the circuit will also be added $n$ times and if the offset voltage value is large, the maximum number of samples that can be added together ($n$) before the circuit saturates, will be limited.

The output differential offset voltage is caused by the mismatches between the circuit's components in the two single-ended signal paths that constitute the differential signal. The most important causes of the offset voltage are the mismatch errors between the transistors $M_1$ in the input differential pair and the mismatch errors between the load transistors $M_3$. The mismatch errors between the two output capacitors ($C_1$) are not important because these capacitors are first charged by a negative current ($I_B$) and then by a positive current ($I_{B2}$), the value of this last current is adjusted by the common mode feedback circuit so that the two currents inject a common mode net charge equal to zero into the capacitors, thus producing a zero offset voltage.

The differential offset voltage created by the mismatch between the two transistors $M_1$ occurs because the two drain currents are not equal due to differences between the threshold voltages $V_T$, $K_N$, $W$ and $L$ parameters of two transistors. The variance of the relative error between the drain current of two equal transistors is given by [37]

$$\sigma^2\left(\frac{\Delta I_D^2}{I_D^2}\right) = \frac{1}{W_1 \cdot L_1} \cdot \left[A_\beta^2 + \frac{4 \cdot A_{VT}^2}{V_{GS1} - V_T}\right] \tag{3.89}$$

where $A_\beta^2$ and $A_{VT}^2$ are statistical parameters dependent of the technology ( $A_\beta^2$ .is related to the size and $K_N$ mismatches between the two transistors and $A_{VT}^2$ is related to the threshold voltage mismatches between the two transistors). The variance of the differential offset voltage caused by the drain current mismatch is given by

$$\sigma^2\left(V_{offset}^2\right) = \frac{1}{W_1 \cdot L_1} \cdot \left[A_\beta^2 + \frac{4 \cdot A_{VT}^2}{V_{GS1} - V_T}\right] \cdot \left(\frac{I_B \cdot \tau}{2 \cdot C_1}\right)^2 \tag{3.90}$$

Using transistors from a 0.18 μm CMOS technology with a $V_{Dsat}$ voltage equal to 100 mV, a $I_B$ value equal to 1 mA, $\tau$=1 ns and a channel length equal to 0.18 μm, results that the differential pair has an output offset with a standard deviation around 1.5 mV.

The mismatch between the two transistors $M_3$ causes a current error between the two drain currents, whose variance is given by an expression similar to expression 3.89. The variance of the differential offset caused by the transistors $M_3$ is also given by an expression similar to expression 3.90, the only difference is the value of the current ($I_{B2}$ instead of $I_B$) and the value of the time interval ( $\tau_2$ instead of $\tau$). The standard deviation value of the differential offset voltage due to the mismatch between the transistors $M_3$ is around 1.3 mV.

The variance of the output offset voltage is the sum of the variances of the offset voltages due to $M_1$ and $M_3$. This means that standard deviation value of the output offset ($\sigma_{offset}$) is around 2 mV. If 100 samples are added together, the offset voltage is also added 100 times, meaning that the maximum accumulated offset voltage can be as high as $100\times3\times\sigma_{offset}$ which, in this case, corresponds to a maximum differential offset voltage of 600 mV.

The maximum output voltage swing of this circuit is limited by the fact that there are five transistors connected in series in the output branch $M_B$, $M_1$, $M_{S2}$, $M_{S3}$ and $M_3$. Each of these transistors (except $M_{S3}$ which is in the linear region), must have a $V_{DS}$ voltage larger than its $V_{Dsat}$ voltage. The single ended voltage swing is also limited by the common mode voltage step ($\Delta V_{CMax}$) that occurs each time a sample is produced. Therefore the maximum peak to peak single-ended output voltage swing of the circuit without significant distortion in the output signal is given by

$$\Delta V_{C_m^p} = V_{DD} - V_{DsatMB} - V_{DsatM1} - V_{DsatM3} - \Delta V_{CM\max} - V_{tol} \tag{3.91}$$

where $V_{tol}$ is a tolerance voltage. For a 1.5 V minimum power supply voltage value, the maximum single ended output voltage swing value is around 0.3 V, which corresponds to a maximum peak to peak differential output voltage swing of 0.6 V. Since the maximum differential output differential voltage swing value is similar to the expected maximum output differential offset value, there is only a very small voltage swing available for the actual signal.

In order to improve the maximum differential output voltage swing, it is necessary to use a fully-differential amplifier with its inputs connected to $v_{Cp}$ and $v_{Cm}$, as shown in Fig. 3.30. The amplifier has shunt-shunt negative differential feedback path constituted by capacitors $C_2$. This feedback, reduces the input impedance of the amplifier, thus if the amplifier had infinite gain, the input impedance of the amplifier would be zero, resulting that ideally the differential voltage $v_{out1}$ would be zero.

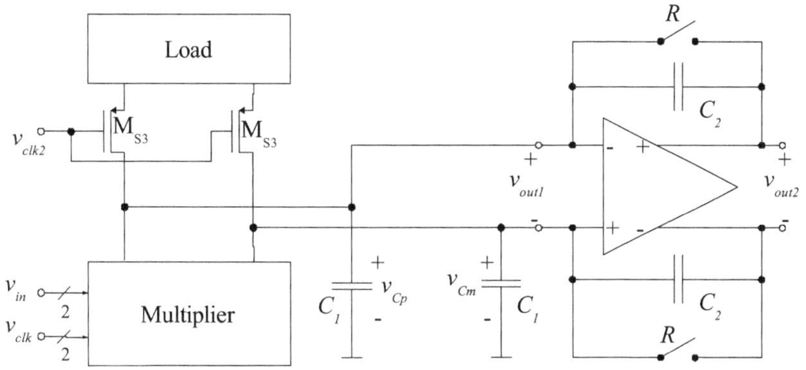


**Fig. 3.30** Switched-integrator with a fully-differential amplifier to increase the maximum differential output voltage swing and provide a low impedance output voltage source

The amplifier differential output voltage can be calculated considering that the differential charge injected into the capacitors $C_1$ is completely transferred to the capacitors $C_2$, producing the output differential voltage. This amplifier works essentially in the same way as the amplifier and capacitor $C_2$ in the circuit depicted in Fig. 3.14. The amplifier differential output voltage is given by[12]

$$v_{out2} = v_{out1} \cdot \frac{C_1}{C_2} \tag{3.92}$$

The previous expression shows that the signals present in the output of the amplifier, can be calculated directly from the signals present in the output of the switched-integrator. The output noise power and sampled signal can be calculated simply by substituting $v_{out}$ in expression 3.92 with the noise voltage expression (3.67, 3.71 or 3.85) or with the sampled voltage expression (expression 3.56). This amplifier allows to introduce an extra voltage gain in the signal path ($C_1/C_2$) with a low penalty in the input referred noise, if sample averaging is used (see Section 3.3.5).

If a two stage amplifier topology, with an output stage constituted by two transistors ($M_N$ and $M_P$) is used, the maximum peak to peak differential voltage swing is maximized and is given by

$$\Delta V_{out2\,\max} = 2 \times \left(V_{DD} - V_{DsatMN} - V_{DsatMP} - V_{tol}\right) \tag{3.93}$$

Comparing the previous expression with expression 3.92, it is clear that there is an improvement the maximum differential output voltage swing of the circuit. However, if a larger number of samples is added, the output of the amplifier will eventually saturate. Using an amplifier with two stages is also useful to buffer the

[12] This is assuming that an ideal amplifier with infinite DC gain is used.

output voltage. The amplifier has low output impedance, while the output impedance of the switched-integrator must be, necessarily, very high.

In order to reduce the differential offset voltage, it is necessary to introduce a calibration cycle, where the differential offset voltage is subtracted from the output differential voltage. This calibration cycle must be repeated each time a sample is added to the output voltage, resulting in a operation mode where the circuit performs a sampling cycle followed by a calibration cycle for each sample of the input signal added to the output voltage. In order to subtract the differential offset voltage from the output voltage, it is necessary to invert the polarity of the differential current signals coming from the load circuit and the differential switch and to disconnect the input signal (otherwise the input signal would also be canceled). This is achieved by adding additional switches to the circuit a pair of switches ($M_{S3a}$) is added in parallel with the switches $M_{S3}$ to invert the polarity of the current $I_{B2}$. A second pair of switches ($M_{S2a}$) is added in parallel with switches $M_{S2}$ to invert the polarity of $I_B$ and finally a switch ($M_{S4}$) is placed in parallel with the input signal to shunt the input signal, significantly attenuating the input voltage. This transistor is designed in order to have an ON resistance value at least 10 times smaller than the source resistance. The schematic of the switched-integrator circuit with the calibration switches (shaded in gray) is depicted in Fig. 3.31.

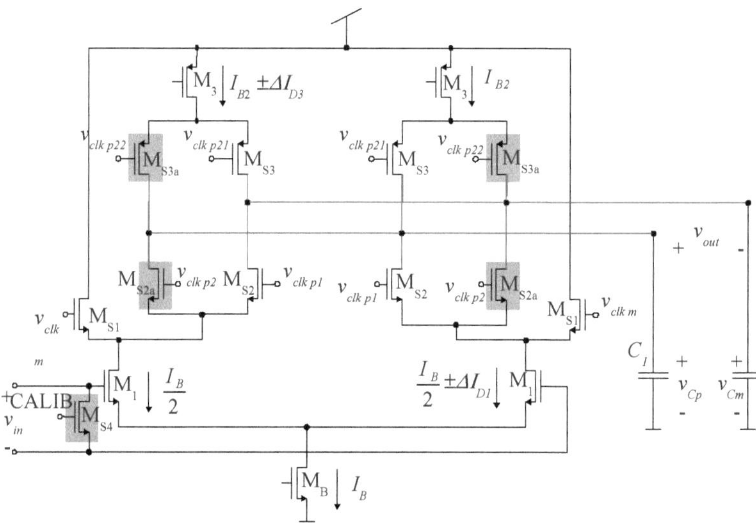


**Fig. 3.31** Switched-integrator and load circuits, with the additional offset calibration switches in gray. These switches are controlled by the digital signal CALIB

The calibration procedure can be demonstrated analyzing the previous circuit and considering (as is illustrated) that due to mismatch errors the drain current of the transistor $M_1$ on the right side is $\Delta I_{D1}$ higher and that the drain current of the transistor M3 on the left side is $\Delta I_{D3}$ higher. In this case, at the end of the sampling cycle the differential output voltage would be equal to

$$\Delta V_{out1} = \frac{\Delta I_{D1} \cdot \tau}{C_1} - \frac{\Delta I_{D3} \cdot \tau_2}{C_1} + \frac{v_{in} \cdot gm \cdot \tau}{C_1} \tag{3.94}$$

where $v_{in}$ is the amplitude of the input rectangular pulse. After the calibration cycle the differential output voltage would be equal to

$$\Delta V_{out1} = \frac{\Delta I_{D1} \cdot \tau}{C_1} - \frac{\Delta I_{D3} \cdot \tau_2}{C_1} + \frac{v_{in} \cdot gm \cdot \tau}{C_1} - \frac{\Delta I_{D1} \cdot \tau}{C_1} + \frac{\Delta I_{D3} \cdot \tau_2}{C_1} = \frac{v_{in} \cdot gm \cdot \tau}{C_1} \tag{3.95}$$

and thus the offset is canceled.

The extra switches are activated during the calibration cycle, controlled by the digital CMOS signal CALIB. When this signal is activated the calibration switches are selected, instead of the regular switches ($M_{S3a}$ instead of $M_{S3}$ and $M_{S2a}$ instead of $M_{S2}$). The actual time instant of when the switches open or close, is still controlled by the original clock signals, $v_{clk}$ for $M_{S2a}$ and $v_{clk2}$ for $M_{S3a}$. The CALIB controls an analog multiplexer that selects which transistor receives the original clock signal and which transistor remains OFF. The signal $v_{clk2}$ is split into two signals $v_{clk21}$ (active when CALIB is "0") and $v_{clk22}$ (active when CALIB is "1"). The signal $v_{clkp}$ is split into two signals $v_{clkp1}$ and $v_{clkp2}$, since the $v_{clkp}$ is part of a differential clock signal ($v_{clk}$) and therefore the new clock signals $v_{clkp1}$ and $v_{clkp2}$ must be symmetrical in relation to the clock signal $v_{clkm}$. The new timing diagram of the circuit with the calibration cycle is shown in Fig. 3.32.

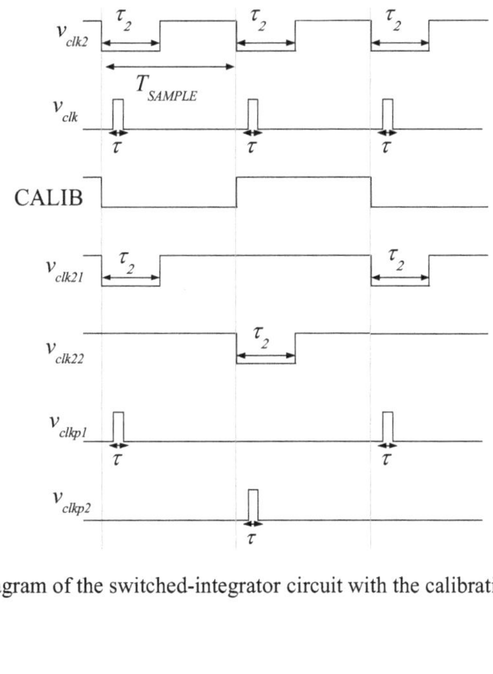


**Fig. 3.32** Timing diagram of the switched-integrator circuit with the calibration cycle

The signals in the previous timing diagram are all generated from a master clock signal (the *PRF* clock that controls the pulse transmission), whose rising edges are shown in the timing diagram by the vertical gray lines. The sampling clock is a delayed version of this master clock. The amount of delay is controlled by the programmable delay circuit.

### 3.3.10 Differential Switched Integrator Noise Analysis

The differential output noise power of the differential switched-integrator circuit is mainly due to the contributions of noise sources from the input differential pair transistors ($M_1$), the load transistors ($M_3$) and the source resistance ($R_s$). The contributions from the noise sources in the switch transistors ($M_{S1}$, $M_{S2}$ and $M_{S3}$) to the output noise can be neglected because these are very small when compared to the contributions from the other noise sources. A simplified schematic of the differential switch circuit containing the relevant noise sources (in gray), is shown in Fig.3.33. It is important to note that applying a differential AC voltage signal to the differential pair, results that the voltage $v_p$ will remain constant and can be treated as a virtual ground for AC signals.

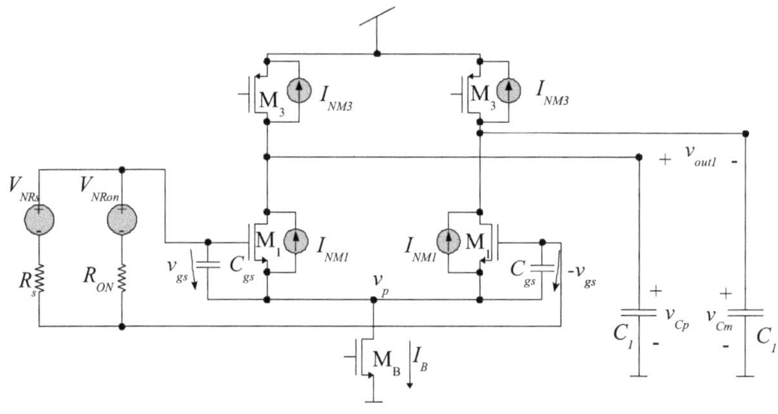


**Fig. 3.33** Simplified schematic of the differential integrator circuit with the relevant noise sources in gray

When the offset calibration is used, each output sample produced by the switched integrator has a differential output noise voltage due to the noise sources present during the sampling phase and due to the noise sources present during the calibration phase. The noise power added by the noise sources in the transistors $M_1$ and $M_3$ is the same during the sampling and offset calibration phases. The noise power added by the ON resistance of the $M_{S4}$ switch and from the source resistance is different during the sampling and offset calibration phases. The expressions of the output differential noise power due to the first two noise sources (transistors $M_1$ and $M_3$) have already been calculated (see expressions

3.85 and 3.86). The differential output noise due to the source resistance, during the sampling phase (switch $M_{S4}$ is OFF), can be calculated following the methodology described in Section 3.3.7. The low pass circuit formed by $R_s$ and the two $C_{gs}$ capacitances (see Fig. 3.33), filters the PSD of the thermal noise due to $R_s$, resulting in a noise voltage across the gate of the transistor $M_1$, $v_{gs}$, with a PSD equal to

$$V_{gs\,N\,R_S}^2 \Delta f = 4 \cdot K_B \cdot T \cdot R_S \cdot \left| \frac{1}{1 + C_{gs} \cdot R_S \cdot j \cdot 2 \cdot \pi \cdot f} \right|^2 \tag{3.96}$$

The voltage $v_{gs}$ is converted into a current by each of the transistor transconductance in the differential pair and these currents are applied to each of the $C_1$ capacitance when the switch is closed. Therefore, the noise power present at the output, due to the resistor $R_s$ can be calculated using

$$\begin{aligned} V_{N\,R_S}^2 &= \int_0^{+\infty} 4 \cdot K_B \cdot T \cdot R_S \cdot \left| \frac{1}{1 + C_{gs} \cdot R_S \cdot j \cdot 2 \cdot \pi \cdot f} \right|^2 \cdot \left| \frac{\sin(\pi \cdot f \cdot \tau)}{C_1 \cdot \pi \cdot f} \right|^2 df = \\ &= \frac{2 \cdot K_B \cdot T \cdot gm^2 \cdot R_S}{C_1^2} \cdot \left( \tau + \frac{C_{gs} \cdot R_S}{2} \cdot \left( e^{-2 \cdot \tau / C_{gs} \cdot R_S} - 1 \right) \right) \end{aligned} \tag{3.97}$$

The value of the output differential noise power from the source resistance, during the calibration cycle, needs to be recalculated, because of the ON resistance of switch $M_{S4}$. When this switch is ON, the noise voltage from the source resistance is attenuated by the resistive divider formed by the source resistance and the ON resistance of $M_{S4}$. The differential output noise voltage (at the terminals of the capacitors $C_1$) due to this resistance can be calculated changing expression 3.71 to include the resistive divider, resulting in the following expression

$$\begin{aligned} V_{N\,R_S\,2}^2 &= \\ &= \int_0^{+\infty} 4 \cdot K_B \cdot T \cdot R_S \cdot \left| \frac{R_{ON\,S4}}{R_{ON\,S4} + R_S} \right|^2 \left| \frac{2 \cdot gm}{2 + C_{gs} \cdot R_S \| R_{ON\,S4} \cdot j \cdot 2 \cdot \pi \cdot f} \right|^2 \left| \frac{\sin(\pi \cdot f \cdot \tau)}{C_1 \cdot \pi \cdot f} \right|^2 df = \\ &= \frac{2 \cdot K_B \cdot T \cdot gm^2 \cdot R_S}{C_1^2} \cdot \left( \frac{R_{ON\,S4}}{R_{ON\,S4} + R_S} \right)^2 \left( \tau + \frac{C_{gs} \cdot R_S \| R_{ON\,S4}}{2} \cdot \left( e^{-2 \cdot \tau / C_{gs} \cdot R_S \| R_{ON\,S4}} - 1 \right) \right) \end{aligned} \tag{3.98}$$

Since the ON resistance value of the transistor $M_{S4}$ is designed to be at least 10 times smaller than the source resistance, the differential output noise voltage due the source resistance, is much smaller during the offset calibration cycle than during the normal sampling cycle.

The value of the output differential noise power (at the terminals of the capacitors $C_1$) from the ON resistance of transistor $M_{S4}$ can be calculated using the previous formula with some minor changes, resulting in

$$V_{N\,R_{on}}^2 =$$

$$= \int_0^{+\infty} 4 \cdot K_B \cdot T \cdot R_{on\,S4} \cdot \left| \frac{R_S}{R_{ON\,S4} + R_S} \right|^2 \left| \frac{2 \cdot gm}{2 + C_{gs} \cdot R_S \| R_{ON\,S4} \cdot j \cdot 2 \cdot \pi \cdot f} \right|^2 \left| \frac{\sin(\pi \cdot f \cdot \tau)}{C_1 \cdot \pi \cdot f} \right|^2 df =$$

$$= \frac{2 \cdot K_B \cdot T \cdot gm^2 \cdot R_{ON\,S4}}{C_1^2} \cdot \left( \frac{R_S}{R_{ON\,S4} + R_S} \right)^2 \left( \tau + \frac{C_{gs} \cdot R_S \| R_{ON\,S4}}{2} \cdot \left( e^{-2 \cdot \tau / C_{gs} \cdot R_S \| R_{ON\,S4}} - 1 \right) \right) \tag{3.99}$$

The total output differential noise power (at the terminals of the capacitors $C_1$) is the sum of the two previous output differential noise powers for the calibration cycle, plus the output differential noise power from the sampling cycle. The resulting differential output noise power is approximately two times larger than the noise power of the circuit during the sampling cycle. The expression of the differential output noise after a sampling and calibration cycles is given by

$$V_{Nout}^2 = 2 \times V_{N\,M1}^2 + 2 \times V_{N\,M3}^2 + V_{N\,R_S}^2 + V_{N\,R_S\,2}^2 + V_{N\,R_{on}}^2 \tag{3.100}$$

The input referred noise voltage of the circuit is obtained dividing the square root of the total output differential noise power by the circuit differential gain:

$$V_{N\,in} = \frac{V_{Nout}}{G_d} \tag{3.101}$$

where the differential gain of the circuit is defined from the input to the terminals of the capacitors $C_1$ and is given by:

$$G_d = \frac{gm}{2 \cdot C_1} \cdot \left( 2 \cdot \tau + C_{gs} \cdot R_S \cdot \left( e^{\frac{-2 \cdot \tau}{C_{gs} \cdot R_S}} - 1 \right) \right) \tag{3.102}$$

note that the differential gain is approximately equal to $gm/C_1$ when the value of $R_s.C_{gs}$ is much smaller than the value of $\tau$. It is also important to note that the importance of the circuit input capacitance is reduced by a factor of 2. This is because the capacitors $C_{gs}$ of the differential pair transistors ($M_1$) are in series.

In Fig. 3.34, the input referred noise value is plotted for $I_B$ changing from 0.1 mA to 100 mA and for three different $V_{Dsat}$ values (100 mV, 120 mV and 200 mV) of the transistor $M_1$ (half of the input differential pair). The temperature value is equal to 290 Kelvin and the $R_s$ value is equal to 50 Ω. The value of $R_{on}$ is equal to 5 Ω.

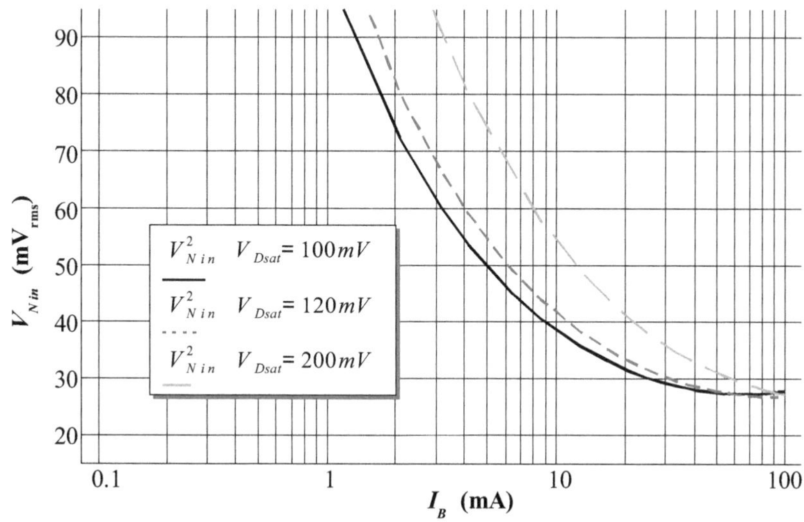


**Fig. 3.34** Input referred noise of the switched integrator circuit, with offset calibration, for different values of $I_B$ and $V_{Dsat}$. The input signal is a rectangular pulse with τ=1 ns, the $R_S$ value is 50 Ω

The previous graph shows that the minimum input referred noise voltage of the circuit is around 27 μVrms and is obtained with a bias current value around 60 mA. The minimum input referred noise value of the differential circuit is obtained for a bias current 6 times larger than the bias current of the simple circuit depicted in Fig. 3.21, this is because the real circuit has an offset calibration phase and has extra noise sources (such as the transistors $M_3$) that where not included in the noise analysis of the simplified switched integrator circuit.

The previous noise analysis does not take into account the noise contribution from the operational amplifier and does not consider the noise reduction that can be obtained using the averaging technique (adding several output samples together). A complete noise analysis of the receiver channel will be realized in Chapter 5. The required noise level for the amplifier will also be determined during the design of the prototype integrated circuit.

## 3.4 Receiver Specifications

The objective of the UWB radar system is to work as a low power, short range, and high resolution sensor. The pulse transmitter of the radar sensor will abide by the limitations imposed by the FCC to imaging UWB systems. Therefore the transmitted pulse must have an EIRP inferior to –41.3 dBm/MHz with a bandwidth inferior to 960 MHz (see Section 2.1). This restriction limits the maximum power of the transmitted pulse and the minimum pulse width.

The first restriction, limits the radar sensor to be only able to distinguish two targets separated by at least 30 cm, this is because the radiated pulse duration must be larger than (approximately) 1 ns in order to have a bandwidth smaller than 960 MHz. The second restriction, limits the maximum range of the radar, since the maximum power of the radiated pulse must be low, the amplitude of the received echoes will be small and if this amplitude is smaller than the receiver noise power the echoes will be undetected.

In the analysis realized in Section 3.2, it was possible to conclude that a metal sphere with a radius equal to 5 cm produces an echo signal with a maximum amplitude smaller than 10 μVrms, when located at 15 m range (see Fig. 3.6) and illuminated by a Gaussian with the maximum allowed power. Smaller targets, or targets constituted by materials other than metal, or targets with different shapes, can produce even smaller maximum echo amplitudes. There is an optimal design for the switched integrator circuit that results in the minimum possible value for the input received noise (see Fig. 3.34). The input referred noise of the receiver channel depends also on the number of samples added together but this number is limited by the offset of the switched-integrator circuit. Taking all these factors into consideration, the minimum input referred noise value of the receiver channel can be as low a 0.85 μVrms, if the switched integrator circuit dissipated around 60 mA and if 1000 samples were added together (which could be difficult because of the offset problem). In this case, the area of the switched integrator circuit would also be very large because it would be necessary to use large values for the $C_1$ capacitors in order to reduce the common mode voltage drop due to such a high bias current.

Due to cost reasons, the prototype IC has an available maximum area of $1.5 \times 1.5\,\mathrm{mm}^2$ (including pads). This area limits the maximum size of the receiver circuit and therefore limits the minimum input referred noise of the circuit. The switched-integrator circuit will be designed to have an input referred noise inferior to 10 μVrms after 100 samples are added together. This value of input referred noise for the receiver should still be enough to detect medium sized targets at a 15 m range. In order to detect a target located at this range it is necessary to introduce a delay into the sample clock signal of 100 ns. As discussed in Section 3.3.2, the sub-sampling technique requires a minimum equivalent sampling frequency larger than twice the maximum input signal bandwidth (approximately 1 GHz). In this case the equivalent sampling frequency is set to 5 times the input signal bandwidth, corresponding to a delay step of 0.2 ns. Therefore the digitally programmable delay circuit must have a maximum delay of 100 ns and a programming resolution equal to 0.2 ns. Since 100 ns can be subdivided into 500 intervals with 0.2 ns width, the programmable delay must have a programming resolution of 9 bits.

# References


1. Introduction to RADAR systems; Merrill I. Skolnik; McGraw-Hill; ISBN 0-07-066572-9
2. Gresham, I.; Jenkins, A.; Egri, R.; Eswarappa, C.; Kolak, F.; Wohlert, R.; Bennett, J.; Lanteri J.-P; Ultra wide band 24 GHz automotive radar front-end; 2003 IEEE Radio Frequency Integrated Circuits (RFIC) Symposium, 8–10 June 2003 Page(s):505–508
3. Gresham, I.; Jenkins, A.; Egri, R.; Eswarappa, C.; Kinayman, N.; Jain, N.; Anderson, R.; Kolak, F.; Wohlert, R.; Bawell, S.P.; Bennett, J.; Lanteri, J.-P.; Ultra-wideband radar sensors for short-range vehicular applications; IEEE Transactions on Microwave Theory and Techniques, Volume 52, Issue 9, Part 1, Sept. 2004 Page(s):2105–2122
4. Kwo, W.C.; Huei, W.; Shreve, G.; Harrison, J.G.; Core, M.; Paxton, A.; Yu, M.; Chen, C.H.; Dow, G.S.; Forward-looking automotive radar using a W-band single-chip transceiver; IEEE Transactions on Microwave Theory and Techniques, Volume 43, Issue 7, Part 1–2, July 1995 Page(s):1659–1668
5. Woll, J.D.; VORAD collision warning radar; Record of the IEEE 1995 International Radar Conference, 1995., 8–11 May 1995 Page(s):369–372
6. Lowbridge, P.L.; Brigginshaw, P.M.; Kumar, B.; Millimetre wave technology for collision avoidance and cruise control; Eighth International Conference on Automotive Electronics, 1991., 28–31 Oct 1991 Page(s):150–154
7. Bennett, H.S.; Brederlow, R.; Costa, J.C.; Cottrell, P.E.; Huang, W.M.; Immorlica, A.A., Jr.; Mueller, J.-E.; Racanelli, M.; Shichijo, H.; Weitzel, C.E.; Bin Zhao; Device and technology evolution for Si-based RF integrated circuits; IEEE Transactions on Electron Devices, Volume 52, Issue 7, July 2005 Page(s):1235–1258
8. Morifuji, E.; Momose, H.S.; Ohguro, T.; Yoshitomi, T.; Kimijima, H.; Matsuoka, F.; Kinugawa, M.; Katsumata, Y.; Iwai, H.; Future perspective and scaling down roadmap for RF CMOS; 1999 Symposium on VLSI Technology, 1999. Digest of Technical Papers. 14-16 June 1999 Page(s):163 - 164
9. Turin, G.; An introduction to matched filters; IEEE Transactions on Information Theory, Volume 6, Issue 3, Jun. 1960 Page(s):311–329
10. Fowler, C.; Entzminger, J.; Corum, J.; Assessment of ultra-wideband (UWB) technology; IEEE Aerospace and Electronic Systems Magazine, Volume 5, Issue 11, Nov. 1990 Page(s):45–49
11. Harmuth, H.F.; Radar equation for nonsinusoidal waves; IEEE Transactions on Electromagnetic Compatibility, Volume 31, Issue 2, May 1989 Page(s):138–147
12. IEEE standard definitions of terms for antennas; IEEE Std 145-1983 22 June 1983
13. Crispin, J.W., Jr.; Maffett, A.L.; Radar cross-section estimation for simple shapes; Proceedings of the IEEE Volume 53, Issue 8, Aug. 1965 Page(s):833–848
14. Hestilow, T.J.; Simple formulas for the calculation of the average physical optics RCS of a cylinder and a flat plate over a symmetric window around broadside; IEEE Antennas and Propagation Magazine, Volume 42, Issue 5, Oct. 2000 Page(s):48–52
15. Colpitts, B.G.; Boiteau, G.; Harmonic radar transceiver design: miniature tags for insect tracking; IEEE Transactions on Antennas and Propagation, Volume 52, Issue 11, Nov. 2004 Page(s):2825–2832
16. Dybdal, R.B.; Radar cross-section measurements; Proceedings of the IEEE, Volume 75, Issue 4, April 1987 Page(s):498–516
17. Behzad Razavi; RF Microelectronics; Prentice Hall, ISBN 0-13-887571-5
18. A. Bruce Carlson; Communication Systems; McGraw Hill, ISBN 0-07-100560-9
19. Won Namgoong; Lerdworatawee, J.; Noise figure of digital communication receivers-revisited; IEEE Transactions on Circuits and Systems I, Volume 51, Issue 7, July 2004 Page(s):1330–1335
20. XYZs of oscilloscopes; Tektronix; Application Note

21. James D. Taylor and Thomas E. McEwan; The Micropower Impulse Radar; Chapter 6 of "Ultra-wideband Radar Technology", Edited by James D Taylor; CRC Press; ISBN: 0849342678
22. Acharya, A.; Hurst, P.J.; Lewis, S.H.; Thermal noise from switches in a switched-capacitor gain stage; Southwest Symposium on Mixed-Signal Design, 2003. 23–25 Feb. 2003 Page(s):121–126
23. Lee, F.S.; Wentzloff, D.D.; Chandrakasan, A.P.; An ultra-wideband baseband front-end; Digest of Papers 2004 IEEE Radio Frequency Integrated Circuits (RFIC) Symposium, 6–8 June 2004 Page(s):493–496
24. Shelton, R.; Adkins, A.; Noise Bandwidth of Common Filters; IEEE Transactions on Communications, Volume 18, Issue 6, Dec. 1970 Page(s):828–830
25. Shaeffer, D.K.; Lee, T.H.; A 1.5-V, 1.5-GHz CMOS low noise amplifier; IEEE Journal of Solid-State Circuits, Volume 32, Issue 5, May 1997 Page(s):745–759
26. Ismail, A.; Abidi, A.A.; A 3-10-GHz low-noise amplifier with wideband LC-ladder matching network; IEEE Journal of Solid-State Circuits, Volume 39, Issue 12, Dec. 2004 Page(s):2269–227
27. Lerdworatawee, J.; Won Namgoong; Wide-band CMOS cascode low-noise amplifier design based on source degeneration topology; IEEE Transactions on Circuits and Systems I: Regular Papers, Volume 52, Issue 11, Nov. 2005 Page(s):2327–2334
28. Paulino, N.; Rebelo, H.; Pires, F.; Ventim Neves, I.; Goes, J.; Steiger-Garcao, A.; Design of a spiral-mode microstrip antenna and matching circuitry for ultra-wide-band receivers; IEEE International Symposium on Circuits and Systems, 2002. ISCAS 2002. Volume 3, 26–29 May 2002 Page(s):875–878
29. Heydari, P.; Lin, D.; A performance optimized CMOS distributed LNA for UWB receivers; Proceedings of the IEEE 2005 Custom Integrated Circuits Conference, 2005, 18–21 Sept. 2005 Page(s):330–333
30. Bruccoleri, F.; Klumperink, E.A.M.; Nauta, B.; Wide-band CMOS low-noise amplifier exploiting thermal noise canceling; IEEE Journal of Solid-State Circuits, Volume 39, Issue 2, Feb. 2004 Page(s):275–282
31. Scholten, A.J.; Tromp, H.J.; Tiemeijer, L.F.; Van Langevelde, R.; Havens, R.J.; De Vreede, P.W.H.; Roes, R.F.M.; Woerlee, P.H.; Montree, A.H.; Klaassen, D.B.M.; Accurate thermal noise model for deep-submicron CMOS; 1999. IEDM Technical Digest. International Electron Devices Meeting, 5–8 Dec. 1999 Page(s):155–158
32. Kwangseok Han; Kwyro Lee; Hyungcheol Shin; Thermal noise modeling for short-channel MOSFETs; International Conference on Simulation of Semiconductor Processes and Devices, 2003. SISPAD 2003. 3–5 Sept. 2003 Page(s):79–82.
33. Claasen, T.; Mecklenbrauker, W.; On stationary linear time-varying systems; IEEE Transactions on Circuits and Systems, Volume 29, Issue 3, Mar 1982 Page(s):169–184
34. Bastos, J.; Steyaert, M.; Sansen, W.; A high yield 12-bit 250-MS/s CMOS D/A converter; Proceedings of the IEEE 1996, Custom Integrated Circuits Conference, 1996. 5–8 May 1996 Page(s):431–434
35. Chi-Hung Lin; Bult, K.; A 10-b, 500-MSample/s CMOS DAC in 0.6 mm2; IEEE Journal of Solid-State Circuits, Volume 33, Issue 12, Dec. 1998 Page(s):1948–1958
36. Gilbert, B.; A precise four-quadrant multiplier with subnanosecond response; IEEE Journal of Solid-State Circuits, Volume 3, Issue 4, Dec. 1968 Page(s):365–373
37. Pelgrom, M.J.M.; Duinmaijer, A.C.J.; Welbers, A.P.G.; Matching properties of MOS transistors; IEEE Journal of Solid-State Circuits, Volume 24, Issue 5, Oct 1989 Page(s):1433–1439

# Chapter 4
# Digitally Programmable Delay

**Abstract** this chapter deals with the problem of generating a clock signal with a programmable delay. This clock is necessary in the radar system to define the sampling instant in the receiver channel. The delay in this clock signal is relative to the transmit clock signal and is used to determine the target distance. A new digitally programmable delay architecture that can have a large programming linearity is proposed. This architecture is based on a digital ΣΔ modulator that controls a 1-bit digital to time converter, whose output is filtered by a delay lock-loop, thus producing the delayed clock signal. This architecture is analyzed and designed, at high level, to meet the required specifications for the radar system. A high level model is used to simulate the behavior of this circuit and validate the design. The electronic sub-blocks necessary to build this circuit are described, analyzed and design methodology for each sub-block is derived. These circuits are implemented using differential clock signals in order to reduce the noise level in the radar system.

The objective of the digitally programmable delay is to produce a variable delay between the transmitted clock and the received clock. This delay determines the distance at which an echo can be detected. The maximum RADAR range is determined by the maximum delay and the range sweeping step is defined by the minimum delay step. The programmable delay will have a maximum delay equal to 100 ns, corresponding to a maximum radar range of 15 m, and a delay programming step smaller than 0.2 ns, corresponding to a ranging resolution smaller than 3 cm.

Although it is easy to create a variable delay in a clock signal, by adjusting the bias current of a current-starved inverter for example, it is important that the delay value is digitally programmable, in order to facilitate the operation of the radar system.

Traditionally, digitally programmable delay lines are realized using a cascade of delay elements and selecting the output of the element corresponding to the desired delay. Typically this delay line is inserted into a delay-locked-loop (DLL) to guaranty that the delay is not affected by process and temperature variations. This type of delay line suffers from element mismatch, resulting in limited delay resolution [1]. The programming resolution is important because, when doing a range sweep, it is necessary to guaranty that each programming code of the delay corresponds to a precise distance. Another problem, is that if both a small delay step and a large maximum delay are required, the delay line will be composed of a large number of delay elements [2], [3], each one having a delay equal to the delay step, in this case this approach would result in at least 256 delay elements (8-bit resolution). Normally, the delay in a circuit is inversely proportional to the circuit



N. Paulino et al., *Low Power UWB CMOS Radar Sensors*, 115–166.

current, therefore this solution (using a large number of small delay elements) would result in a large power dissipation and in large area.

The selected architecture [4] for the digitally programmable delay is capable of generating a digitally programmable delay whose resolution is not affected by element mismatch and is only limited by the response time to a new programming code.

## 4.1 Architecture

A Delay Lock Loop (DLL), such as the one depicted in Fig. 4.1, is a feedback system, where the delay produced by a voltage controlled delay line (VCDL) into a clock signal, is adjusted until is equal to one or more periods of the reference clock [5], [6]. This loop works in a similar way to a Phase Locked Loop (PLL), except that the input frequency value is always the same as the output frequency value. The loop compares the phase (delay) of the reference clock with the phase of the delayed clock using a phase detector (PD), that, in turn, controls a charge pump (CP). The output of the CP goes through a low-pass filter and the resulting voltage is used to control the delay in the VCDL. The loop will stop adjusting the control voltage when the rising (or the falling) edges of both the reference and delayed clocks coincide. Normally the input clock of the VCDL is the same clock signal used as the reference clock.

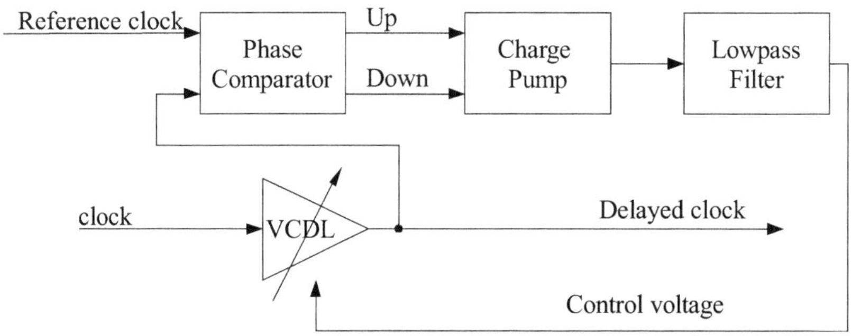


**Fig. 4.1** Block diagram of a Delay Locked Loop

If two different clock signals with the same frequency, are used as the reference clock and the input clock for the VCDL (as shown in Fig. 4.1), the loop will try to adjust the amount of delay in the VCDL to be equal to the average delay between the two clock signals, so that the rising edges of the delayed clock and reference clock signals coincide (on average). The reference clock does not need to be a clean signal (i.e. with low jitter noise) because the DLL will act as a low pass filter that attenuates the excess jitter noise from the clock signals. The jitter noise in the output clock will be due to the jitter of both the reference clock and the input clock of the VCDL (master clock), the master clock jitter noise can be very small, if a crystal oscillator [7] is used as source for this signal.

The noisy reference clock is generated switching between the master clock and a delayed version of this clock (with a fixed delay equal to $T_D$), as shown in Fig. 4.2. A digital ΣΔ modulator, with a fixed input digital code, controls the switching sequence, this switching produces a reference clock with a large jitter noise, but with an average delay value controlled by the digital input code of the ΣΔ modulator. To avoid spikes in this reference clock, the delay $T_D$ should be inferior to half period of the master clock.

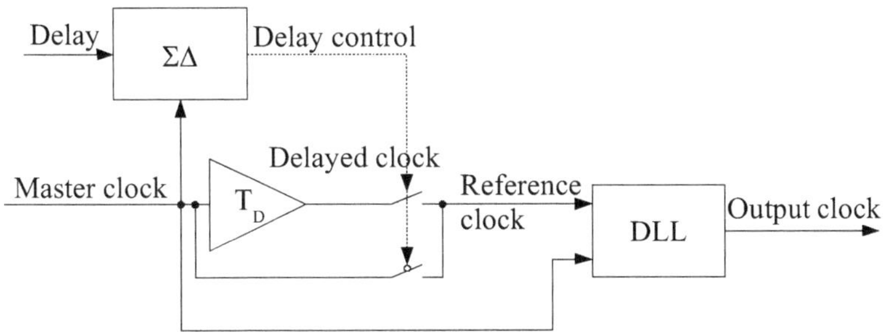


**Fig. 4.2** Architecture of the digitally programmable DLL

This reference clock is applied to the DLL and the loop will adjust the delay produced by the VCDL in the master clock to match the average delay of the reference clock. The low-pass characteristic of the DLL will filter most of the jitter noise from the output clock. To illustrate the operation of the circuit the waveforms of these signals can be seen in Fig. 4.3 for the case of 50% desired delay.

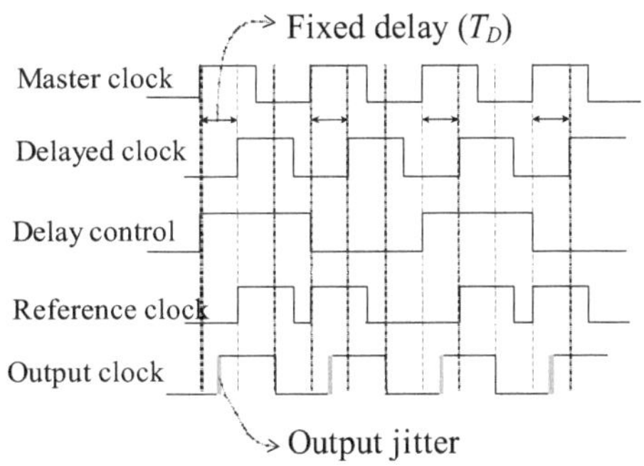


**Fig. 4.3** Signal Waveforms in the programmable DLL for 50% delay

In this figure, the delay control signal is ON 50% of time. This signal controls the switching of the reference clock when it is OFF the reference clock is equal to the master clock, when it is ON the reference clock is equal to the master clock delayed by $T_D$. The resulting reference clock has an average delay value of $0.5.T_D$. The reference clock is applied to the DLL that adjusts the delay of the VCDL until it is equal to the average value of the reference clock. The resulting DLL output clock, as shown in the figure, has a delay of $0.5.T_D$ with some remaining jitter noise.

A precise fixed delay $T_D$ can be obtained using the clock scheme shown in Fig. 4.4. The master clock period is set to $4.T_D$ and this master clock is obtained dividing by 2 a system clock (clk in Fig. 4.4), the delayed clock is obtained dividing by 2 the inverted system clock. The delay between the rising edges of these clock is $T_D$. The reference clock is generated from these two clock signals as before, using the delay control signal to switch between the two.

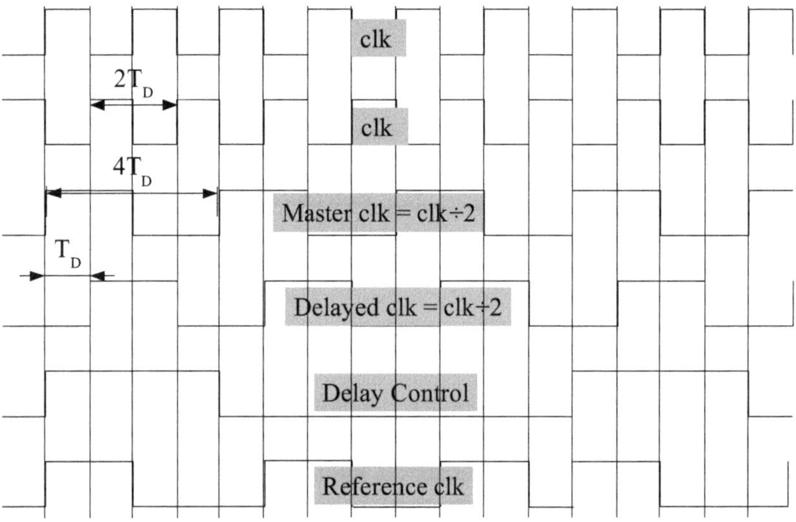


**Fig. 4.4** Generation of the delayed clock using a system clock with a period equal to $2T_D$

This process for generating the delay $T_D$ is very accurate; the error in the value of $T_D$ is very small if a system clock with a good accuracy is used. The main error sources for the $T_D$ value are the delays introduced by the gates in the clocks signal path (the master clock and delayed clock have different signals paths). These delays can be very small when compared to the value of $T_D$ and both clock paths can be designed to have almost equal delays.

## 4.2 High Level Analysis and Design

The circuit of Fig. 4.2 can be considered as a digital-to-analog converter (where the analog output signal is a time-delay, instead of a voltage or a current) with one bit resolution. The analogy between a "normal" digital to analog converter (DAC) and this digital to time converter (DTC) implies that the DTC circuit can be analyzed similarly to the way a DAC circuit is analyzed. When the control signal is ON the delay is $T_D$, when this signal is OFF the delay is 0. This means that the "*lsb*" of this 1 bit DTC is equal to $T_D$ and that the spectral power density of the quantization noise of this DTC, can be approximated by [8]

$$S(z) = \frac{T_D}{12} \times \frac{1}{F_{CLK}} \times \left| NTF(z) \right|^2 \tag{4.1}$$

where *NTF*(*z*) is the noise transfer function of the ΣΔ modulator, for a modulator of order *n* is given by [8]

$$NTF(z) = \left(1 - z^{-1}\right)^n \tag{4.2}$$

The *NTF*(*z*) is a high pass filter since it has *n* zeros located at DC. This means that the ΣΔ modulator attenuates the quantization noise at low frequencies and amplifies the quantization noise at high frequencies. This means that the jitter noise in the reference clock is mainly in frequencies close to the clock frequency. The DLL loop acts as low-pass filter and reduces most of the quantization noise energy. It should be noted that, the ΣΔ, depending on the input DC level, can produce tones that concentrate the energy of the quantization noise below the cut-off frequency of the DLL filter and thus the real output jitter noise is higher than the predicted by these simple expressions. These tones can be reduced if a dither signal is applied to the ΣΔ [8]. The minimum order of the DLL is related to the order of the ΣΔ, however the DLL must be a second-order system because of closed loop stability problems. The optimum order of the ΣΔ modulator can be determined calculating the *rms* value of the jitter noise at the output for different orders of ΣΔ modulators. The variance of the jitter noise at the output of the DLL is obtained calculating the output power of the quantization noise, integrating the spectral noise density of the input jitter multiplied by the squared modulus of the DLL transfer function, according to [9]

$$\sigma_{jitter}^2 = \int_{-F_{clk}/2}^{F_{clk}/2} S\left(e^{j \cdot 2 \cdot \pi \cdot f \cdot T_{clk}}\right) \cdot \left| H_{DLL}\left(e^{j \cdot 2 \cdot \pi \cdot f \cdot T_{clk}}\right) \right|^2 df \tag{4.3}$$

where $H_{DLL}$ is the loop second order low pass transfer function with complex poles. The resolution of the programmable delay can be defined (using the analogy to DAC) comparing the *lsb* value with the output quantization noise as shown next

$$res = \log_2 \left[ \frac{T_D}{\sigma_{jitter} \cdot \sqrt{12}} \right] \tag{4.4}$$

Considering that $T_D$=100 ns and that the loop transfer function is a second order low-pass filter with a complex pole frequency $f_p$ and a pole quality factor $Q_p$ equal to 0.8, the dependence of the resolution with the closed loop frequency ($f_p$) is shown in Fig. 4.5 for different orders of ΣΔ modulators (assuming a clock frequency of 2.5 MHz).

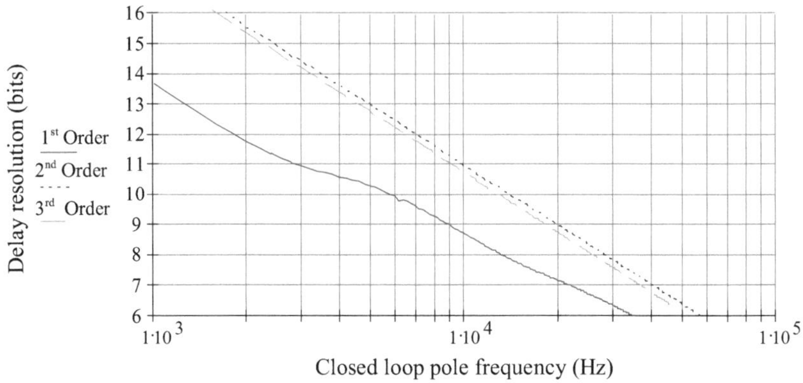


**Fig. 4.5** System resolution (in bits) of the DLL for different closed loop frequencies and a ΣΔ modulator order of 1,2 and 3

From this graph it is possible to conclude that a second order modulator produces the best resolution for a given $f_p$ value (in the case of a second order DLL). It should be noted that there is a trade-off between the $f_p/F_{clk}$ ratio and the system resolution; i.e., a higher resolution means a lower ratio, which translates into a higher settling time for a given $F_{clk}$. The closed loop pole quality factor ($Q_p$) does not affect this graph as long as its value is between 0.5 and 1.5. As an example a resolution of 9 bits, corresponding to an *rms* jitter of 52 ps, is obtained with $f_p$=20 kHz.

At this point it is important to clarify the meaning of the programming resolution. A DAC (or a DTC as it is the case here) based on a ΣΔ modulator differs from a traditional DAC in an important point there are no pre-assigned output levels. In theory, any output value can be selected by using the correct programming code at the input of the ΣΔ modulator. Of course, each programming code corresponds to an average output value, each output value having an error with a variance value given by expression (4.3).

The previous analysis shows that the closed-loop transfer function should have a certain pole frequency to obtain a given resolution. In order to design the loop to have the desired pole frequency (and quality factor), it is necessary to determine the relation between the closed-loop parameters ($f_p$ and $Q_p$) and the values of each loop element.

The DLL transfer function can be calculated assuming that the loop is under lock condition. In that case the difference between the reference clock and the delayed clock is necessarily small and an incremental analysis of the loop can be done. An incremental analysis linearizes the loop using the derivatives of the different transfers' functions of the loop sub-blocks. All the DC levels in the loop variables are eliminated.

The phase detector samples the delay between the two clock signals and activates the charge pump for the duration of the sampled delay ($d[n]$) [10]. The sample frequency is, of course $F_{clk}$. An example of the waveforms for the phase comparator is shown in Fig. 4.6; in this figure the inputs are signals $A$ and $B$ where signal $B$ defines the sampling instant of the phase-delay between the two signals. The gain ($K_{TD}$) of the phase detector is 1 because the active time of the $UP$ and $DOWN$ signals is the same as the sampled delay time duration. The behavior of the phase detector will be analyzed in more detail in Section 4.3.3.

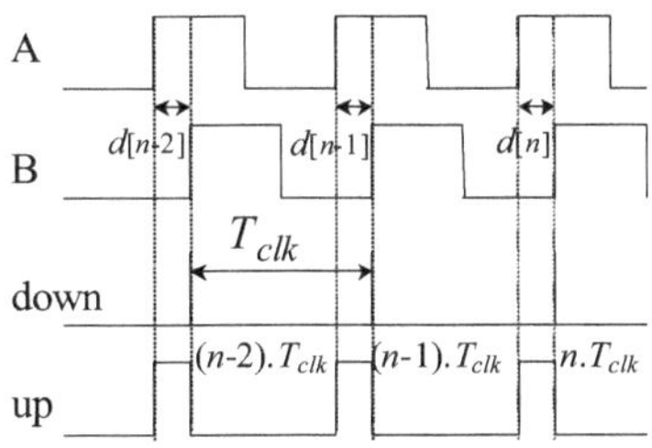


**Fig. 4.6** Typical waveforms in a phase detector

The sampled delay ($d[n]$) controls the time duration that a current source inside the charge pump will be ON (with a constant current value equal to $I_p$). This current will charge a capacitor ($C$), increasing or decreasing its voltage, depending on which current source is on, which is controlled by the "Up" and "Down" signals. A simplified schematic of the charge pump is shown in Fig. 4.7.

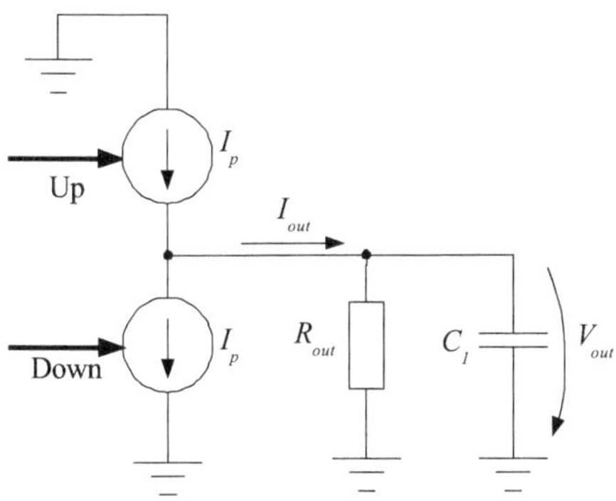


**Fig. 4.7** Schematic of the simplified Charge pump circuit

This circuit is a first order system and therefore its output voltage transient behavior is determined by a damped exponential. The output voltage varies with time according to

$$v_{out}(t) = v_{out}\left[(n-1)\cdot T_{clk}\right]\cdot e^{\frac{-t}{R_{off}\cdot C_1}} \approx v_{out}\left[(n-1)\cdot T_{clk}\right]$$
$$\left[(n-1)\cdot T_{clk}\right] \le t \le \left[n\cdot T_{clk} - d(n)\right] \tag{4.5}$$

$$v_{out}(t) = v_{out}\left[(n-1)\cdot T_{clk}\right] + \left(1 - e^{\frac{-t}{R_{out}\cdot C_1}}\right)\cdot\left(I_p\cdot R_{out} - v_{out}\left[(n-1)\cdot T_{clk}\right]\right)$$
$$\left[n\cdot T_{clk} - d(n)\right] \le t \le \left[n\cdot T_{clk}\right] \tag{4.6}$$

Equation (4.5) represents the capacitor voltage when the current source is OFF. This voltage can be considered approximately constant, because the OFF resistance of the current source ($R_{off}$) is very high and therefore, the decay of the output voltage during $T_{clk}$ can be considered negligible. Equation (4.6) represents the capacitor voltage when the current source is ON. The exponential function can be simplified using $e^x \approx 1 + x$ (if $x$<<1), in this case $T_{clk}$ is very small compared to $R_{out}.C_1$ (where $R_{out}$ is the ON output resistance of the current source) and so equation (4.6) can be simplified to

$$v_{out}(t) \approx v_{out}\left[(n-1)\cdot T_{clk}\right]\cdot\left(1 - \frac{t}{R_{out}\cdot C_1}\right) + \frac{I_p}{C_1}\cdot t$$
$$\left[n\cdot T_{clk} - d(n)\right] \le t \le \left[n\cdot T_{clk}\right] \tag{4.7}$$

Since the behavior of the circuit is completely described by knowing the output voltage at $t = n\cdot T_{clk}$ the output voltage can be calculated using a discrete time equation. Assuming that $T_{clk} / R_{out}\cdot C_1 << 1$ is valid, the resulting discrete time equation is

$$v_{out}[n] = v_{out}[n-1] + d[n]\cdot\frac{I_p}{C_1} \tag{4.8}$$

To obtain a second-order loop, it is necessary to add a RC circuit to the loop to create another pole. This RC circuit is decoupled from the charge pump output using a transconductance circuit (that also provides a voltage gain $G$). Assuming that this extra pole has a time constant equal to $\tau_{RC} = R_2\cdot C_2 >> T_{clk}$ , the output voltage at the end of a clock period can be obtained by a similar analysis and is given by

$$v_o[n] = v_o[n-1]\cdot\left(1 - \frac{T_{clk}}{\tau_{RC}}\right) + G\cdot v_{out}[n-1]\cdot\frac{T_{clk}}{\tau_{RC}} \tag{4.9}$$

The combined transfer function is obtained applying the Z transform to equations (4.8) and (4.9) and solving in order of $V_o(z)$ and is given by (the gain of the transfer function is [V/s])

$$H_1(z) = \frac{V_o(z)}{D(z)} = \frac{G \cdot \frac{I_P}{C_1} \cdot \frac{T_{clk}}{\tau_{RC}} \cdot z^{-1}}{\left(1 - z^{-1}\right) \cdot \left(1 - z^{-1} \cdot \left(1 - \frac{T_{clk}}{\tau_{RC}}\right)\right)} \tag{4.10}$$

The voltage controlled delay line produces a delay that depends on the control voltage (the amount of delay obtained for a given control voltage is determined by the gain factor $K_{VCDL}$), this dependence is not instantaneous, i.e., after the control voltage changes the amount of delay introduced changes to the new value after a certain time. A good approach for the dynamic of the control voltage in the VCDL is to consider a first order system. If, in reality there are more poles, they must be located at frequencies much higher than the dominant pole in order to guaranty a stable circuit. Using this model and assuming that the control voltage changes only at $n.T_{clk}$,[1] the output delay is given by

$$d_{out}(t) = K_{VCDL} \cdot v_o\left[(n-1) \cdot T_{clk}\right] \cdot e^{-p \cdot t} + K_{VCDL} \cdot v_o\left[(n-1) \cdot T_{clk}\right] \cdot \left(1 - e^{-p \cdot t}\right) \tag{4.11}$$

where $p$ is the dominant pole inside the VCDL, if the VCDL is designed to make $p.T_{clk} >> 1$ then equation (4.11) simplifies to the discrete time equation

$$d_{out}[n] = K_{VCDL} \cdot v_o[n] \tag{4.12}$$

The linearized discrete time model of the DLL loop is shown in Fig. 4.8

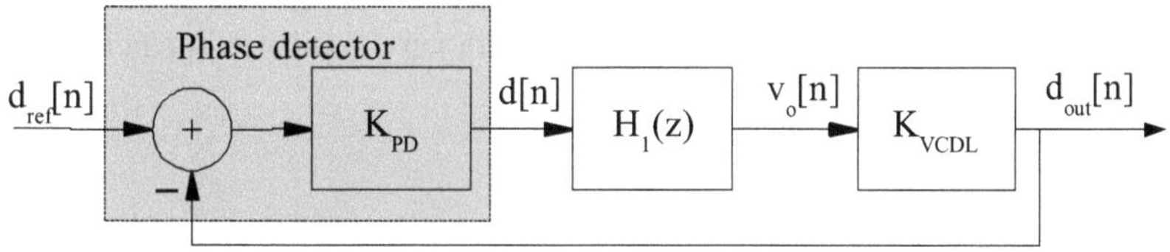


**Fig. 4.8** Linearized discrete time model of the DLL loop

The DLL closed loop transfer function is obtained by applying Mason's rule resulting in

---

[1] In reality the voltage $v_o(t)$ changes throughout $T_{CLK}$, but the worst case for the settling time of the VCDL happens when $v_o(t)$ changes in one step at a discrete time instant.

$$H_{CL}(z)=\frac{K_{PD}\cdot K_{VCDL}\cdot H_1(z)}{1+K_{PD}\cdot K_{VCDL}\cdot H_1(z)}=$$
$$=\frac{K_{PD}\cdot K_{VCDL}\cdot G\cdot I_P\cdot T_{clk}\cdot z}{K_{PD}\cdot K_{VCDL}\cdot G\cdot I_P\cdot T_{clk}+C_1\cdot\tau_{RC}\cdot(z-1)\cdot\left(z-1+\frac{T_{clk}}{\tau_{RC}}\right)} \tag{4.13}$$

In this expression $K_{PD}$ is the gain of the phase detector and $K_{VCDL}$ is the gain of the VCDL. The frequency response of this transfer function is evaluated using $z=e^{j\cdot 2\cdot\pi\cdot f\cdot T_{clk}}$, but if $f_p$ is small compared to $F_{clk}$ (as is the case), the frequency response can be evaluated using $z=1+j\cdot 2\cdot\pi\cdot f\cdot T_{clk}$. Using this approximation in (4.13) results in a continuous time second order transfer function (prototype function) from which the pole quality factor and frequency expressions are obtained

$$f_p=\frac{1}{2\cdot\pi}\cdot\sqrt{K_{PD}\cdot K_{VCDL}\cdot G\cdot\frac{I_P}{C_1}\cdot\frac{T_{clk}}{\tau_{RC}}} \tag{4.14}$$

$$Q_p=\frac{\sqrt{K_{PD}\cdot K_{VCDL}\cdot G\cdot\frac{I_P}{C_1}\cdot\frac{T_{clk}}{\tau_{RC}}}}{\frac{1}{\tau_{RC}}+K_{PD}\cdot K_{VCDL}\cdot G\cdot\frac{I_P}{C_1}\cdot\frac{T_{clk}}{\tau_{RC}}} \tag{4.15}$$

These expressions are used to determine the values of $\tau_{RC}$ and $I_p/C_1$ that result in the desired $f_p$ and $Q_p$ values. Since the previous values are determined using expressions that correspond to a prototype continuous function and the DLL transfer function is discrete, it is expected that the DLL discrete time transfer function behavior differs from the prototype continuous time transfer function behavior. The frequency responses of both transfer functions are very similar as it can be seen in Fig. 4.9.

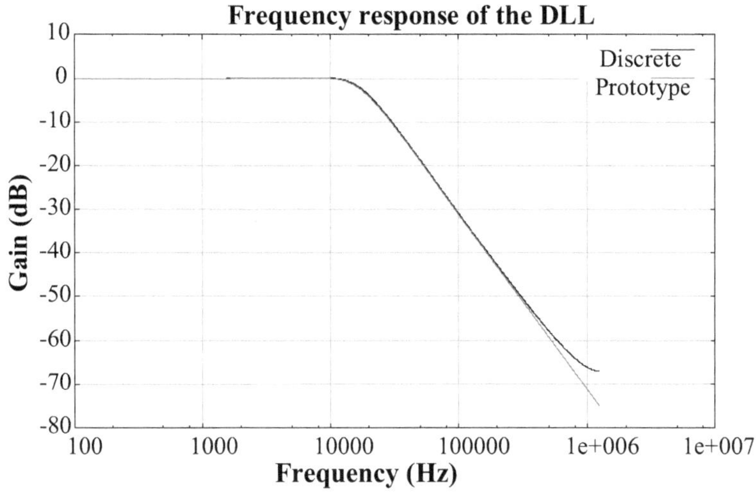


**Fig. 4.9** Closed-loop frequency response, calculated using the exact discrete transfer function and the prototype continuous time transfer function

The closed loop DLL frequency response is evaluated using expression (4.13), this frequency response has a small difference to the prototype frequency response at low frequencies (as expected from the analysis), but at frequencies close to $F_{CLK}/2$ it diverges. The fact that the attenuation of the DLL transfer function is smaller than the prototype transfer function at high frequencies causes an increase in the output jitter noise since the input jitter noise is located mainly at high frequencies. Therefore the programming resolution predicted by expression (4.4) is smaller than the values shown in Fig. 4.5.

In order to validate the system design it is necessary to model the block diagram depicted in Fig. 4.2 (including the sigma delta). The DLL is modeled using the linearized block diagram of Fig. 4.8. This model is used to run several transient time simulations for different stimulus. These transient simulations confirm that the output jitter noise is indeed higher and therefore it is necessary to reduce the closed loop pole frequency to achieve a 9 bit programming resolution. The closed loop pole has $f_p$=16.5 kHz and $Q_p$ = 0.79, which corresponds to $\tau_{RC}$ = 7.8 μs and $I_p/C_1$ = 90.9 V/ms. The transient behavior of the output delay of the system was simulated with a 75 ns step applied to the control signal and the system response is compared to the transient behavior of the ideal prototype second order system. These two responses are depicted in Fig. 4.10 and there is a good adjustment between system output and the expected output.

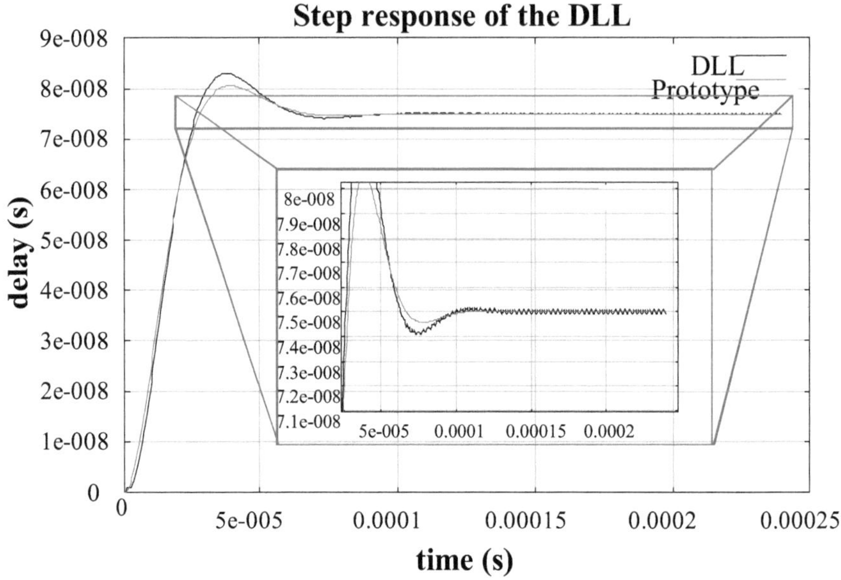


**Fig. 4.10** Step response of the programmable DLL compared with the ideal system for a 75 ns final value. The insert graphic is a magnification of the end value of the response

The system response is almost identical to the prototype response. The magnification of the end value shows that the output delay is not constant and has jitter noise (due to the quantization noise of the sigma delta).

To verify the programming range a stair case input is applied to the control input and the output delay measured, this is shown in Fig.4.11.

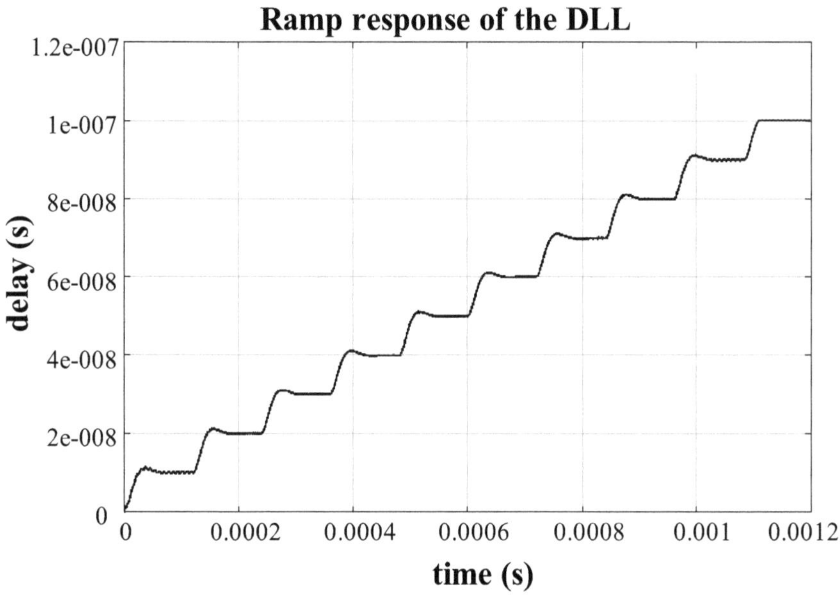


**Fig. 4.11** Output delay of the DLL for a ramp in the control delay

Several simulations were done where different delays were selected. After the response of the system stabilizes, the real delay, the *rms* and peak-to-peak value of the jitter noise are measured. These results can be observed in Fig.4.12.

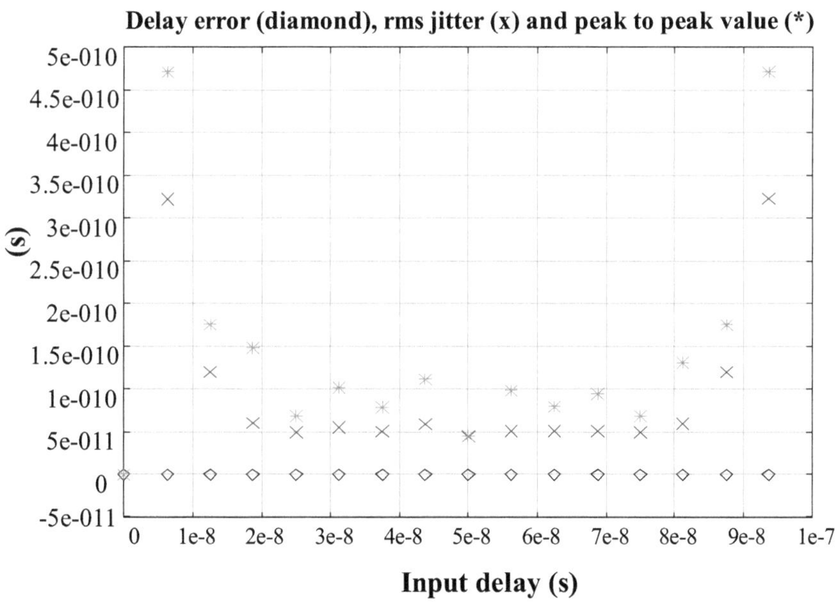


**Fig. 4.12** Simulated output delay error values (diamond), rms jitter value (X) and peak-to-peak error values (*)

From this last graph it is possible to conclude that the output *rms* jitter level depends on the desired delay value, this variation is because the sigma delta noise increases when the input code is close to 1 or –1. The average *rms* jitter noise is 55 ps, which is in accordance with the desired programming resolution (9 bits). The jitter noise value increases both for small and large delays, because the sigma delta modulator begins to saturate and the quantization noise appears as low frequency tones that the DLL low-pass filter functions does not eliminate. This phenomenon is easily explained observing the sigma delta output bit-stream (the delay control signal) in Fig. 4.3, the control signal in this figure is a square wave with a frequency value equal to $F_{CLK}/2$ and corresponds to a delay equal to $0.5.T_D$. If a different delay ($x.T_D$) is desired the control signal will be a sequence of "ones" and "zeros" with an average value equal to $x$. If the desired value of $x$ is close to 1 or to 0, the bit-stream must be constituted by long sequences of "ones" or "zeros" in order to achieve the desired average value.[2] These long sequences have low frequency components that can fall below the DLL closed loop pole frequency and therefore they are not attenuated.

The values of the DLL components vary due to manufacture imperfections. These variations are random and cause the DLL transfer function to also vary randomly. To evaluate the effect of the random variations of the DLL components in the output jitter, it is necessary to realize a Monte Carlo analysis [11]. This analysis is performed by first choosing a random set of values (within the manufacture tolerances) for $R_2$, $C_1$, $C_2$ and $K_{VCDL}$ (the other loop components are not considered in the analysis because the variation of their values is much smaller than these 4). Using these random values the system is simulated and the output jitter is calculated, this process is repeated 10000 times and the histogram of the resulting jitter values, obtained in each simulation is shown in Fig. 4.13.

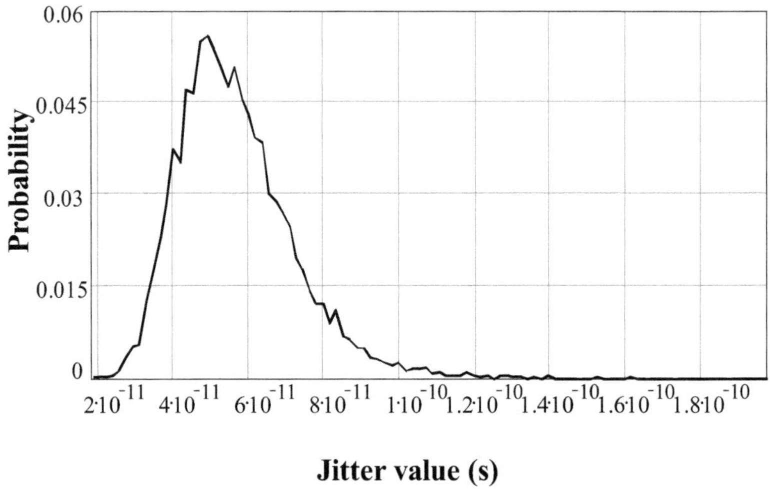


**Fig. 4.13** Output jitter value probability obtained from a 10000 cases Monte-Carlo analysis

[2] This subject is discussed in more detail in Appendix C.

# 4.3 Electrical Design

The design of the circuits that implement the sub-blocks of the digital programmable delay is going to use differential signals when ever possible. Using analog differential circuits reduces the interference from other circuits (and the substrate), increases maximum voltage swing, therefore increasing the dynamic range of the signals. Differential digital circuits have the advantage over CMOS circuits of drawing a constant current from the power supply instead of large amplitude current spikes. This will limit the amount of noise that the digital circuits generate and that can couple into the analog circuits. Next, the designs of the different circuits implementing the sub-blocks are going to be described.

These circuits are going to be designed in a 0.18 μm CMOS technology and will work with a minimum value of 1.5 V power supply and dissipate as little power as possible. The circuits will be designed to work with a junction temperature between –40°C and 100°C.

## *4.3.1 Differential Buffer*

The basic building block circuit for processing differential digital signals is the differential buffer[3] composed by a differential pair and a resistive load. The schematic of a differential buffer with resistive load is shown in Fig. 4.14

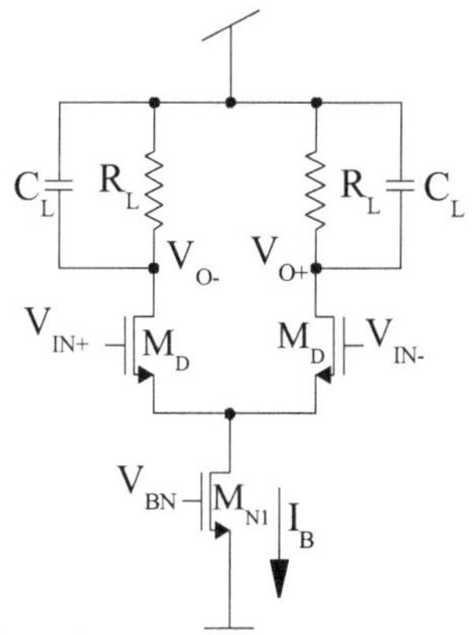


**Fig. 4.14** Differential buffer schematic

This buffer operates with differential digital signals, i.e., signals that have only two valid voltage levels, each representing a digital value. These voltage levels correspond to the buffer two differential output voltage levels when the differential pair is saturated (one of the transistors $M_D$ is ON and the other is OFF). The value of the buffer output voltage swing ($V_{swing}$) is determined by the buffer bias

[3] Or inverter, depending on the polarity of the output differential voltage.

current ($I_B$) and by the load resistance value ($R_L$) simply using Ohm's law ($V_{SWING} = R_L \times I_B$). In order to design this circuit it is necessary to understand how the response of the circuit to a square wave input signal (in terms of delay, rise and fall time) changes with the circuit design parameters. A digital input signal is characterized by a rise time (considered to be equal to the fall time) and an amplitude. The amplitude of the input signal should be large enough to saturate the differential pair and guaranty digital behavior in the circuit. The maximum value of the output amplitude is limited by the power supply voltage ($V_{DD}$) and the $V_{dsat}$ voltage of the transistor $M_{N1}$. The maximum value for the output amplitude must be smaller than $V_{DD}$ - $V_{dsat}$ -$V_{tol}$, where $V_{tol}$ is a safety margin to guaranty that $M_{N1}$ is in the saturation region and is around 200 mV. This results that the maximum buffer voltage swing of the buffer is around 1 V for a power supply value of 1.5 V.

The differential pair produces an output differential current ($\Delta I = I_{d1} - I_{d2}$), which is a non-linear function of the input differential voltage ($\Delta V_{in}$). This function can be calculated using the level 1 MOS model, resulting in the following expression

$$
\begin{aligned}
\Delta I &= -I_B && \text{if} \quad \left|\frac{\Delta V_{in}}{V_{Dsat}}\right| < -\sqrt{2} \\
\Delta I &= \Delta V_{in} \cdot \frac{I_B}{V_{Dsat}} \cdot \sqrt{1 - \frac{1}{4} \cdot \left(\frac{\Delta V_{in}}{V_{Dsat}}\right)^2} && \text{if} \quad -\sqrt{2} < \left|\frac{\Delta V_{in}}{V_{Dsat}}\right| < \sqrt{2} \\
\Delta I &= I_B && \text{if} \quad \left|\frac{\Delta V_{in}}{V_{Dsat}}\right| < \sqrt{2}
\end{aligned}
\qquad (4.16)
$$

The previous expression assumes two distinct operating regions for the differential pair a linear region, where both transistors $M_D$ are ON and a non-linear region or saturation region where one of the transistors is OFF. From this expression it is possible to conclude that an input signal with an amplitude of at least $1.5.V_{Dsat}$ guaranties the saturation of the differential pair and thus produces digital behaviour in the circuit. Unfortunately this expression is too complicated to use in the following analysis, because it is not easy to manipulate expressions with a square root. The effort would be unnecessary, because this expression is obtained using a simple transistor model that does not take into account second order effects, thus resulting in discrepancies between expression (4.16) and the real behaviour of the differential pair. Therefore, a more simple expression, corresponding to the linearisation of expression (4.16) is chosen. This new expression can be adjusted (via the $x$ parameter) to match the differential pair predicted behaviour to the electrical simulations using BSIMv3 transistor models for a given technology. The output differential current as a function of the differential input voltage can be approximated by

$$\Delta I = -I_B \qquad if \quad \left|\frac{\Delta V_{in}}{V_{Dsat}}\right| < -x$$

$$\Delta I = \Delta v_{in} \cdot gm_D \qquad if \quad -x < \left|\frac{\Delta V_{in}}{V_{Dsat}}\right| < x \tag{4.17}$$

$$\Delta I = I_B \qquad if \quad \left|\frac{\Delta V_{in}}{V_{Dsat}}\right| < x$$

where $gm_D = k \cdot I_B / V_{Dsat}$ is the transconductance value of the differential pair transistors ( $k = 1/x$ ).

The two previous expressions are compared to the electrical simulation, for the case of a differential pair with a bias current of 100 μA and a $V_{dsat}$ of 250 mV, in Fig. 4.15. Analyzing this figure it is clear that expression (4.17) is a good approximation to the real differential pair DC transfer function when both transistors in the input differential pair are ON. When one of the transistors starts to turn OFF the discrepancies between expression (4.17) and the simulation increase. The value of $x$ is around 1.66 (this value changes with the process corner) which is slightly larger than $\sqrt{2}$. Analyzing the differential pair DC characteristic in the figure, it is clear that the input signal should have a minimum voltage swing equal to $2 \times V_{Dsat}$, to achieve the complete saturation of the differential pair. But to guaranty digital behavior for the buffer it is better to add a safety margin of at least 10% and use a minimum voltage swing of $2.2 \times V_{Dsat}$.

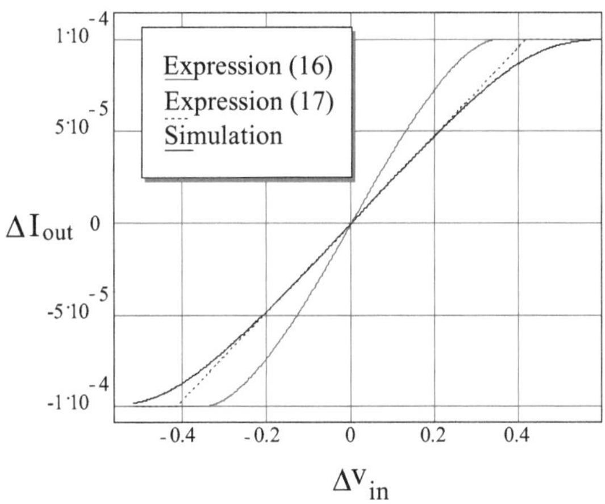


**Fig. 4.15** Differential output current as a function of the differential input voltage of a differential pair with a bias current of 100 uA and a $V_{dsat}$ of 250 mV, obtained from expression 4.16, expression 4.17 with $x$=1.66 and $k$=0.6 and an electrical simulation using the BSIMv3 transistor model

The differential pair exhibits a non-linear behaviour, as was shown in the previous analysis. This non-linearity prevents the calculation of the exact dynamic response of the circuit. However an approximation of the output voltage behaviour can be calculated assuming that the circuit has two distinct operating regions saturation (when one of the transistors of the differential pair is OFF) and linear operation (when both transistors of the differential pair are ON). This assumption corresponds to the behavior described by expression (4.17). The transition between the two operating regions is controlled by the input differential voltage. When $\Delta V_{in} < |x \cdot V_{Dsat}|$ the circuit is in the linear region and both transistors are ON, otherwise the circuit is saturated and one of the transistors is OFF. Considering that the differential input signal is a ramp signal, similar to the one of Fig. 4.16, the circuit will change operating regions at the time instants shown in the figure.

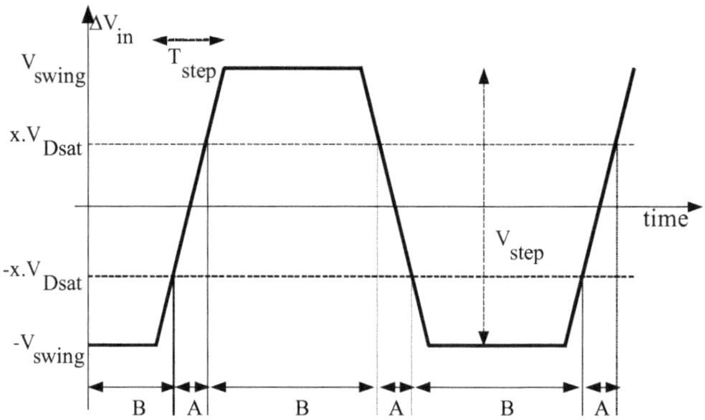


**Fig. 4.16** Typical differential input voltage signal. The circuit is in the linear region inside the A time intervals and saturated inside the B time intervals. The input signal is characterized by a rise and fall time equal to $T_{step}$ and an amplitude equal to $V_{step}$

It is possible to determine the output voltage behavior of the differential buffer for the input signal defined in Fig. 4.16 calculating the output voltage using two incremental circuit models of the buffer circuit, one for each operating region and then selecting the correct output voltage, depending on the time instant. This corresponds to realizing a linear circuit analysis for each time interval (A and B). The time intervals are defined by the transition time points of the differential pair (defined in expression 4.17), which in the case of the first rising edge are given by

$$A = [T_1, T_2] \quad T_1 = \frac{V_{SWING} - x \cdot V_{Dsat}}{V_{STEP}} \cdot T_{STEP} \quad T_2 = \frac{V_{SWING} + x \cdot V_{Dsat}}{V_{STEP}} \cdot T_{STEP} \tag{4.18}$$

When the differential pair is in the saturation region (time interval B) one of the transistors $M_D$ is OFF and all the other transistors are in the saturation region. The small signal analysis is realized using the incremental circuit depicted in Fig. 4.17.

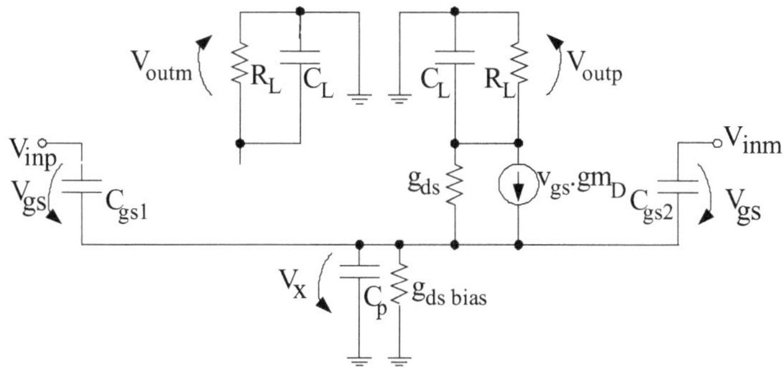


**Fig. 4.17** Incremental circuit model for the differential buffer in the saturation

Applying the KCL to this circuit and solving the resulting equation system for $v_{om}$ and $v_x$ results in the following $s$ transfer function for the output voltage (assuming $g_{ds} << g_{mD}$ and that $C_{gs1} << C_{gs2}$)

$$\frac{\Delta V_o}{\Delta V_{in}} \approx \frac{gm_D \cdot \left(g_{ds\,bias} + s \cdot C_p\right)}{\left(1/R_L + s \cdot C_p\right) \cdot \left(g_{ds\,bias} + s \cdot C_p + s \cdot C_{gs2} + gm_D\right)} \tag{4.19}$$

This expression is similar to the transfer function of a single transistor amplifier, where the voltage gain value is approximately equal to the drain impedance value divided by the source impedance value.

The output voltage in the time domain, can be calculated by applying the inverse *Laplace* transformation to the following expression

$$\Delta V_o(s) = \Delta V_{in}(s) \cdot \left(\frac{\Delta V_o}{\Delta V_{in}}\right) \tag{4.20}$$

where $\Delta V_{in}(s)$ is the *Laplace* transform of the input signal. To obtain a completely linear analysis, the input signal should be defined using the following peace wise linear description (during the first rising edge)

$$\Delta v_{in}(t) = \begin{cases} -V_{SWING} & \textit{if } t > T_{STEP} \\ -V_{SWING} + t \cdot \frac{V_{STEP}}{T_{STEP}} & \textit{if } 0 < t < T_{STEP} \\ -V_{SWING} + t \cdot \frac{V_{STEP}}{T_{STEP}} - \left(t - T_{STEP}\right) \cdot \frac{V_{STEP}}{T_{STEP}} & \textit{if } t > T_{STEP} \end{cases} \tag{4.21}$$

This input signal is constructed adding (or subtracting) ramps and thus its *Laplace* transform is also obtained adding or subtracting the *Laplace* transform of ramps, resulting in

$$\Delta V_{in}(s) = \begin{cases} 0 & \text{if } t > T_{STEP} \\ \frac{1}{s^2} \cdot \frac{V_{STEP}}{T_{STEP}} & \text{if } 0 < t < T_{STEP} \\ \frac{1}{s^2} \cdot \frac{V_{STEP}}{T_{STEP}} - \frac{1}{s^2} \cdot \frac{V_{STEP}}{T_{STEP}} \cdot e^{-s \cdot T_{STEP}} & \text{if } t > T_{STEP} \end{cases} \tag{4.22}$$

Substituting expression (4.22) into expression (4.20) and then calculating the inverse *Laplace* transform results in the expression for the differential output voltage ($v_{o\_sat}(t)$) when the circuit is saturated.

$$v_{o\ sat}(t) = \begin{cases} 0 & \text{if } t < 0 \\ v_{o\ ramp\ sat}(t) & \text{if } 0 < t < T_{STEP} \\ v_{o\ ramp\ sat}(t) - v_{o\ ramp\ sat}(t - T_{STEP}) & \text{if } t > T_{STEP} \end{cases} \tag{4.23}$$

where

$$v_{o\ ramp\ sat}(t) = \frac{V_{STEP}}{T_{STEP}} \cdot \frac{p_1 \cdot p_2}{z} \cdot k \cdot \left[ \frac{e^{-p_1 \cdot t}}{p_1 \cdot (p_1 - p_2)} - \frac{z \cdot e^{-p_1 \cdot t}}{p_1^2 \cdot (p_1 - p_2)} + \frac{e^{-p_2 \cdot t}}{p_2 \cdot (p_1 - p_2)} - \frac{z \cdot e^{-p_2 \cdot t}}{p_2^2 \cdot (p_1 - p_2)} + \frac{1 + z \cdot t}{p_1 \cdot p_2} - \frac{z}{p_1^2 \cdot p_2} - \frac{z}{p_2^2 \cdot p_1} \right] \tag{4.24}$$

and

$$p_1 = \frac{g_L}{C_L} \quad p_2 = \frac{gm_D + g_{ds\ bias}}{C_p + C_{gs1}} \quad z = \frac{g_{ds\ bias}}{C_p} \quad k = \frac{gm_D \cdot g_{ds\ bias}}{g_L \cdot g_{ds\ bias} C_p + gm_D \cdot g_L} \tag{4.25}$$

Equation (4.24) describes the saturated buffer differential output voltage, for a ramp input. This voltage changes only due to the drain current variation of the transistor in the differential pair that is ON. The current of this transistor changes with the gate voltage because the source impedance ($g_{ds\_bias}$ in parallel with $C_p$) is not infinite. Ideally the transistor source voltage should follow the input ramp applied to the gate. Although $g_{ds\_bias}$ can be made very low, by designing the current source transistor with a large channel length, increasing this channel length results in increasing the parasitic capacitance $C_p$. So in a typical buffer design the source impedance is a compromise between the $g_{ds\_bias}$ and $C_p$ values. Analyzing the multiplying factor of equation (4.24), considering that $g_{ds\_bias}$ >> ($g_{mD}$ or $g_L$), results in

$$\frac{p_1 \cdot p_2}{z} \cdot \frac{k}{T_{STEP}} \approx \frac{C_p}{C_L} \cdot \frac{gm_D}{C_p + C_{gs1}} \cdot \frac{1}{T_{STEP}} \tag{4.26}$$

From this expression it is clear that if the load capacitance is larger than the parasitic capacitance and that the source node time constant is smaller than the input ramp time step, the output voltage when the buffer is saturated does not change significantly and expression (4.23) is not important in describing the buffer behavior. In the case of a fast input ramp (small $T_{step}$) expression (4.23) is very important in describing the buffer output voltage behavior.

When the circuit is in the linear region a small signal analysis can be performed, using the circuit depicted in Fig. 4.18, assuming that all the transistors are in the saturation region.

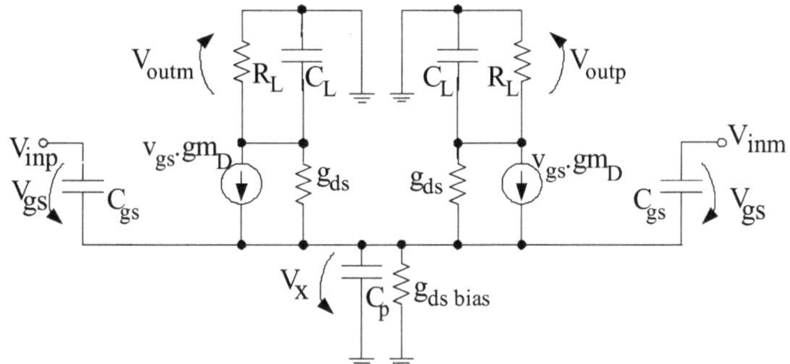


**Fig. 4.18** Incremental circuit model for the differential buffer in the linear region

Applying the KCL to this circuit and solving for $v_{om}$ and $v_{op}$ results in the following $s$ transfer function (assuming that the circuit is symmetrical and that $v_{inp}$=$-v_{inm}$ results that $v_x$=0 )

$$\frac{\Delta V_o}{\Delta V_{in}} = -\frac{gm_D \cdot R_L}{\left(g_L + g_{ds\,bias} + s \cdot C_L\right)} \tag{4.27}$$

where $g_L = 1/R_L$ . The differential output voltage for the linear region is calculated using the same methodology as before, considering that the input signal is defined as

$$\Delta v_{in}(t)=\begin{cases} -V_{Dsat} & if\ t<T_1 \\ -V_{Dsat}+(t-T_1)\cdot\dfrac{V_{STEP}}{T_{STEP}} & if\ T_1<t<T_2 \\ -V_{Dsat}+(t-T_1)\cdot\dfrac{V_{STEP}}{T_{STEP}}-(t-T_2)\cdot\dfrac{V_{STEP}}{T_{STEP}} & if\ t>T_2 \end{cases} \tag{4.28}$$

using this definition for the input signal results in an elegant definition of the saturation of the differential pair. Note that the differential output current is $-I_B$ if $\Delta v_{in}<-V_{Dsat}\cdot x$ and $I_B$ if $\Delta v_{in}>V_{Dsat}\cdot x$. After calculating the inverse *Laplace* transform of (4.27) multiplied by the *Laplace* transform of (4.28) the expression of the output voltage for the linear region[4] is obtained

$$\Delta v_{o\,lin}(t)=\begin{cases} -V_{SWING} & if\ t<T_1 \\ -V_{SWING}+\dfrac{V_{STEP}}{T_{STEP}}\cdot\dfrac{gm}{g_{ds}+g_L}\cdot\left[\tau\cdot\left(e^{\frac{T_1-t}{\tau}}-1\right)+(t-T_1)\right] & if\ T_1<t<T_2 \\ -V_{SWING}+\dfrac{V_{STEP}}{T_{STEP}}\cdot\dfrac{gm}{g_{ds}+g_L}\cdot\left[\tau\cdot e^{-t/\tau}\left(e^{T_1/\tau}-e^{T_2/\tau}\right)+(T_2-T_1)\right] & if\ t>T_2 \end{cases} \tag{4.29}$$

where $\tau=C_L/(g_L+g_{ds})$ is the circuit's time constant, $\tau$ determines the speed of the buffer, a small value of $\tau$ compared to $T_{step}$ , results in a buffer who's output voltage change is faster than the input signal; a large value of $\tau$ compared to $T_{step}$, results in a buffer whose output voltage change is slower than the input signal.

Expression (4.29) ($\Delta v_{o\,lin}(t)$) is added to expression (4.23) ($\Delta v_{o\,sat}(t)$) to obtain the complete expression for the circuit's output voltage ($\Delta v_o(t)$). To confirm the validity of the expression $\Delta v_o(t)$, the predicted output voltage is compared to the buffer output voltage obtained by electrical simulation of a differential buffer with three different input ramp signals the first with a time step of 0.5 ns, the second with a time step of 5 ns and the third with a time step of 50 ns. The plots showing the buffer output voltage for these three input voltages are in Figs. 4.19, 4.20 and 4.21. The buffer load capacitance value is the input capacitance of an equal buffer.

---

[4] Note that expression (4.29) is obtained when the circuit has both transistors ON (thus the differential pair is in the linear region) but also describes part of the behavior of the differential pair in the saturation region when the output current is limited to $I_B$.

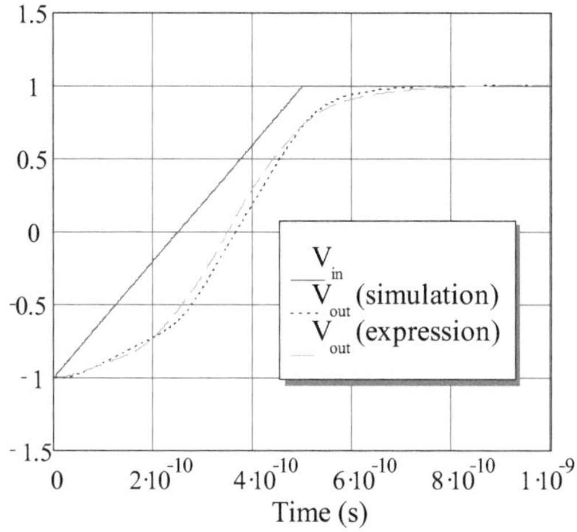


**Fig. 4.19** Plot of the Buffer output voltage for an input ramp signal with step of 0.5 ns

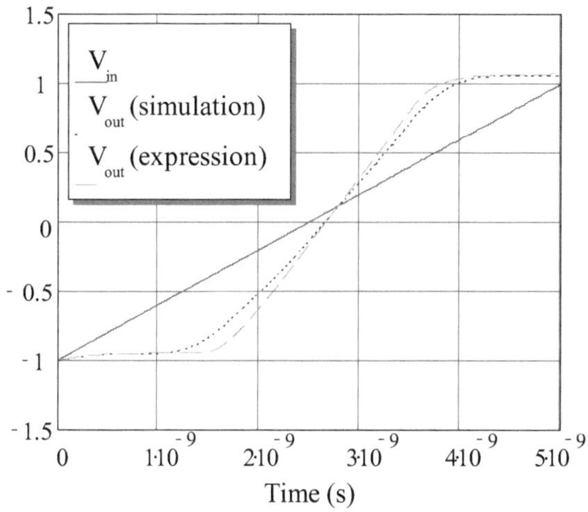


**Fig. 4.20** Plot of the Buffer output voltage for an input ramp signal with step of 5 ns

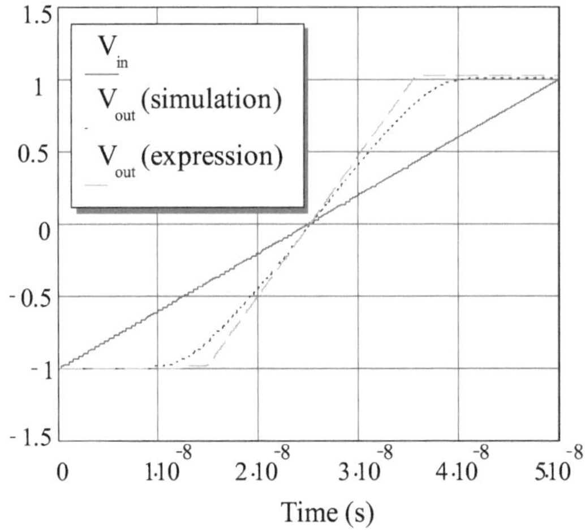


**Fig. 4.21** Plots of the Buffer output voltage for an input ramp signal with step of 50 ns

From the previous plots it is possible to conclude that expressions (4.23) and (4.29) are a good approximation to the buffer output voltage for a ramp input signal. The main difference between the predicted and real values is because of the simplification of the differential pair DC characteristic through expression (4.17). This difference is more visible for a slow input signal such as the one in Fig 4.21. It is also clear that expression (4.23) is only important to describe the output voltage of the buffer for the case of a fast input signal. In Fig. 4.19, the input ramp signal has very small step (0.5 ns) and the output voltage varies significantly from the start of the input ramp, this behavior is described by expression (4.23) fairly accurately. In Fig. 4.20, the input ramp signal has small step (5 ns) and the output voltage varies from the start of the input ramp but with smaller amplitude than in the previous case, it is also visible that there is an overshoot in the output voltage which is also described by expression (4.23). In Fig. 4.21, the input ramp signal has a large step (50 ns) and the output voltage does not change significantly from the start of the input ramp.

In a slow buffer (a desired feature for a delay line) the load capacitance is typically larger than the source parasitic capacitance ($C_L >> C_p$), considering that this buffer is designed to minimize the value of time constant associated to the differential pair source node ($\tau_{source} = \left(C_p + C_{gs}\right)/gm_D$), then expression (4.26) will be much smaller than 1 for input signals with $T_{step} \approx \tau_{source}$. Under these conditions, expression (4.23) is not important to describe the output voltage and this voltage is accurately described using expression (4.29) alone. Expression (4.29) can be used to calculate the time instant *t* at which the output voltage has a certain value. Unfortunately it is not possible to find an exact expression for *t* because expression (4.29) does not admit an explicit solution when solving for *t*. To calculate the delay it is necessary to use numeric methods, such as the Newton method. However it is possible to find a valid approximated expression for the value of this delay for this particular case (a fast signal applied to a slow buffer). Assuming that $T_{step}$ is much smaller than $\tau$, expression (4.29) can be simplified to

$$\Delta v_o(t) \approx -V_{SWING} + \frac{V_{STEP}}{T_{STEP}} \cdot \frac{gm}{g_{ds}+g_L} \cdot \left[ \tau \cdot e^{-t/\tau} \left( e^{T_1/\tau} - e^{T_2/\tau} \right) + (T_2 - T_1) \right] \tag{4.30}$$

This expression can be simplified further considering that $\tau$>>($T_2$ or $T_1$), resulting in

$$\begin{aligned} \Delta v_o(t) &\approx -V_{SWING} + \frac{V_{STEP}}{T_{STEP}} \cdot \frac{gm}{g_{ds}+g_L} \cdot \left[ \tau \cdot e^{-t/\tau} \left( 1 + \frac{T_1}{\tau} - 1 - \frac{T_1}{\tau} \right) + (T_2 - T_1) \right] = \\ &= -V_{SWING} + \frac{V_{STEP}}{T_{STEP}} \cdot \frac{gm}{g_{ds}+g_L} \cdot (T_2 - T_1) \cdot \left( 1 - e^{-t/\tau} \right) \end{aligned} \tag{4.31}$$

Using expression (4.31) it is possible to find the value of time ($t$) when the output voltage is equal to a given voltage value

$$v_o(t) = V_y \Leftrightarrow t = -\tau \times \ln\left( 1 - \frac{V_y + V_{SWING}}{2 \cdot V_{SWING}} \right) \tag{4.32}$$

The delay introduced by the buffer is defined as the time interval between the time instant when the input signal crosses 0 V and the time instant when the buffer output voltage crosses 0 V The buffer delay can be simply calculated as the time it takes the output voltage to change from $-V_{swing}$ to 0 V ($V_y$=0 V) since it is considered that $\tau$>> $T_{step}$. The delay time is given by

$$T_D = -\tau \times \ln\left( 1 - \frac{1}{2} \right) \approx \tau \times 0.7 \tag{4.33}$$

The rise time of the circuit is defined as the time it takes the output voltage to change from 0.1.$V_{step}$ to 0.9. $V_{step}$ (or from –0.8.$V_{swing}$ to 0.8 $V_{swing}$) and is given by

$$T_{rise} = -\tau \times \left[ \ln\left( 1 - \frac{1.8 \cdot V_{SWING}}{2 \cdot V_{SWING}} \right) - \ln\left( 1 - \frac{0.2 \cdot V_{SWING}}{2 \cdot V_{SWING}} \right) \right] \approx \tau \times 2.2 \tag{4.34}$$

The delay time and rise time of a slow buffer with a fast input signal depends only on the value of the circuit time constant ($\tau$). These two expressions are compared to the measured results of the delay time and rise time obtained from the electrical simulation of a buffer with different values of load capacitance (corresponding to different values for $\tau$) in Figs. 4.22 and 4.23. The electrical simulation used an input signal with $T_{step}$ = 0.5 ns and a buffer with $R_L$=10 kΩ. In Fig. 4.22 it is possible to observe that expression (4.33) is accurate, even for low values of $C_L$, in fact the predicted values are accurate for values of the delay down to the $T_{step}$ value (0.5 ns). In Fig. 4.23 it is possible to observe that expression (4.34) is accurate to values of the rise time down to the order of 2 ns. From these figures it is evident that both expressions are very accurate when $\tau$ is much larger than $T_{step}$.

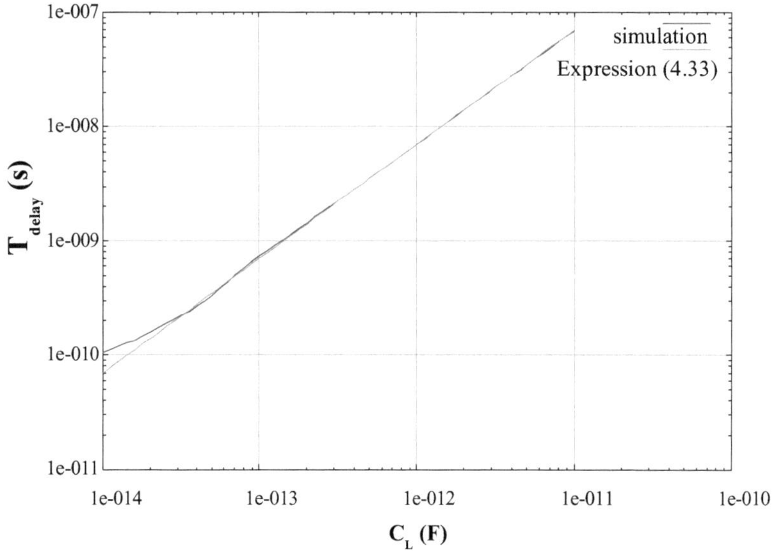


**Fig. 4.22** Plots of the simulated and predicted buffer delays for different values of the load capacitance

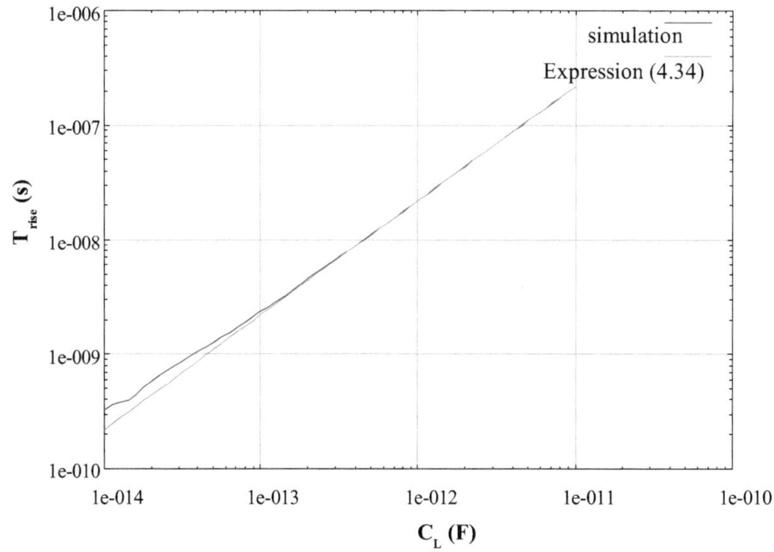


**Fig. 4.23** Plots of the simulated and predicted buffer rise time for different values of the load capacitance

When a slow input signal is applied to a fast buffer it is also possible to find approximated expressions for the delay and rise time of the buffer. This case corresponds to a buffer that is placed at the end of a delay to restore a fast rise time. In this case expression (4.23) has a small contribution to the output voltage and the output voltage can be accurately described using expression (4.29).Considering that $T_{step} >> \tau$, expression (4.29) can be simplified to

$$\Delta v_o(t) \approx -V_{SWING} + \frac{V_{STEP}}{T_{STEP}} \cdot \frac{gm}{g_{ds} + g_L} \cdot \left[\tau \cdot (-1+0) + (T_2 - T_1)\right] \tag{4.35}$$

Solving the previous expression for $t$ results in the following expression

$$v_o(t) = V_y \Leftrightarrow t = \frac{T_{step}}{V_{step}} \times \left[V_{SWING} + x \cdot V_{Dsat} \cdot \frac{V_y}{V_{SWING}}\right] \tag{4.36}$$

Using this expression it is possible to find the delay time (defined as before), given by

$$T_D = \frac{T_{step}}{2} - \frac{T_{step}}{V_{step}} \times \left[V_{SWING} - 0\right] = 0 \tag{4.37}$$

The delay time of a fast inverter is zero. In reality there are no circuits with zero delay. This result is due to the mathematical approximations used in the calculations. This approximation can still be fairly valid, as it is observed in Fig 4.21. In this case the delay of the buffer is negligible compared to the input signal.
The rise time (defined as before) is, given by

$$T_{rise} = \frac{T_{step}}{V_{step}} \times \left(\frac{1}{2} + \frac{1.6 \cdot x \cdot V_{Dsat}}{V_{step}}\right) \tag{4.38}$$

This expression depends on the adjustment parameter $x$, which is an indication that the expression is not very accurate. This can be confirmed observing the difference between the slopes of the simulated output voltage and predicted output voltage in Fig. 4.21. Expression (4.38) can still be used as an indication on how to reduce the rise time of a fast buffer after a slow buffer, design the differential pair transistors of the fast buffer with a $V_{dsat}$ as small as possible. In the extreme (and unrealistic) case of $V_{dsat}$=0 V the minimum rise time would be $T_{step}/2$. From this reasoning it is evident that to obtain a signal with a fast rise time after a slow inverter, it is necessary to use a cascade of several fast inverters.

The previous analysis shows that the speed of a differential buffer depends essentially on the value of $\tau$. For a given value of signal amplitude, the value of $\tau$ depends on the value of the buffer bias current and on the value of the load capacitance. The load capacitance is typically the input capacitance of another differential buffer plus some parasitic capacitance (due to the connecting wires). The value of $\tau$ is therefore

$$\tau = R_L \times C_L = \frac{V_{SWING} \cdot \left(C_{gs} + C_p\right)}{I_B} = \frac{V_{SWING} \cdot C_{gs} \cdot \left(1 + C_p / C_{gs}\right)}{I_B} \tag{4.39}$$

The value of $C_{GS}$ depends on the $V_{dsat}$ and channel length of the differential pair transistors and on the value of the bias current increasing linearly the bias current value while maintaining the $V_{dsat}$ and channel length results in a linear increase of the $C_{GS}$ value. If $C_{GS}$ is much larger than $C_P$, expression (4.38) shows that designing a buffer with a larger bias current (maintaining the channel length and $V_{dsat}$ values) produces almost no decrease in the value of $\tau$. The speed of a buffer, loaded with another buffer of the same size, depends essentially on the channel length and $V_{dsat}$ of the differential pair transistors, i.e., it is related to the $f_T$ of the technology. In a cascade of identical buffers, the ideal bias current, is such, that the value of $C_{GS}$ is about 5–10 times larger than the value of $C_P$.
In conclusion, the design of a differential buffer should use the following rules

- Use a large channel length for the $M_{N1}$ transistor, in order to obtain a small $g_{ds\_bias}$ value when compared to $1/R_L$. A value of 1.5 μm is selected.
- Use a large value for the $V_{dsat}$ of the $M_{N1}$ transistor, to reduce the source parasitic capacitance. The maximum value of $V_{dsat}$ is limited by the voltage swing and the power supply voltage, the $V_{dsat}$ value for the typical process corner is between 250 mV and 350 mV.
- Use minimum channel length for the differential pair transistors to reduce the source node parasitic capacitance.
- Optimize the $V_{dsat}$ value for the differential pair transistors to be as small as possible without increasing too much the source node parasitic capacitance. The value is between 180 mV and 250 mV for the typical process corner.
- The buffer voltage swing should be larger than $2.2 \times V_{Dsat}$, for the worst case of $V_{dsat}$ (in terms of process and temperature variation). An amplitude value of 0.6 V is a good compromise for a 1.5 V power supply.
- The buffer bias current is calculated in order to obtain the desired value of $\tau$. Therefore the bias current is adjusted depending on the buffer load capacitance. Considering the particular case when the load capacitance is equal to the input capacitance of an equal size buffer plus the connection parasitic capacitance, the bias current optimal value is around 100 μA. The load resistance is calculated to obtain the desired voltage swing with the selected bias current value.

## 4.3.2 Replica Bias Circuit

The amplitude of the output voltage ($V_{SWING}$) of the differential buffer depends on the values of the output resistance and of the bias current. The output resistance can vary with process and temperature up to 30% of its nominal value. This variation can be compensated using negative feedback to adjust the bias current ($I_B$) in order to maintain a constant output voltage swing. It would be difficult and inefficient to realize this adjustment inside every differential buffer of the circuit. The feedback circuit would add power dissipation to each buffer and it would be necessary to eliminate the differential signal from the feedback path. It is more efficient to have a replica of the buffer circuit (actually only half of the buffer circuit is needed since one of the outputs is $V_{DD}$) and adjust the output amplitude of this circuit to be equal to the desired value of the voltage swing. The bias current of this half-buffer can then be mirrored to all the buffer circuits [12], [13]. This principle is shown in Fig. 4.24.

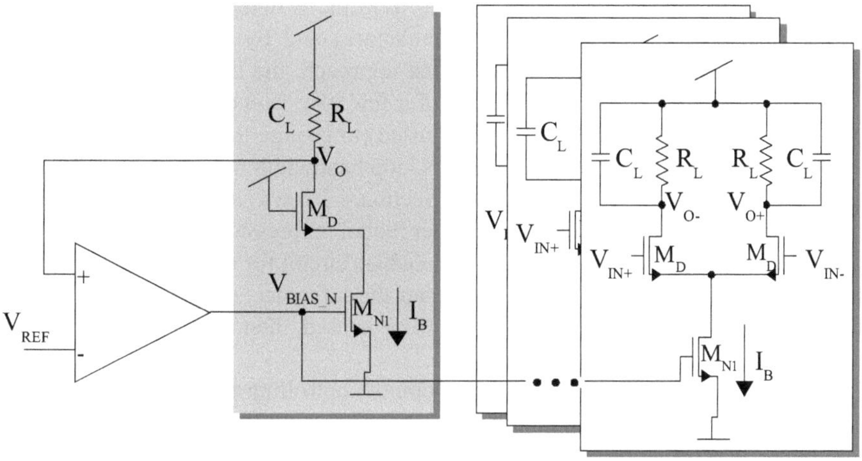


**Fig. 4.24** Replica bias principle

The replica bias circuit works by comparing the reference voltage ($V_{REF}$) with the half-buffer output voltage ($V_o$) using an amplifier. The amplifier output voltage ($V_{BIAS\_N}$) determines the value of the $I_B$ current through transistor $M_{N1}$. The amplifier adjusts the value of this current until the output voltage is almost equal to the reference voltage, therefore $V_{ref}= V_{SWING}$ . The relative error between the $V_{REF}$ voltage and the $V_o$ voltage depends on the feedback loop gain of the replica bias and is given by

$$\frac{V_o}{V_{REF}} = \frac{1}{1 + A \cdot gm_{MN1} \cdot R_L} \tag{4.40}$$

where $A$ is the amplifier gain. This voltage error is not critical, since the amplitude has a safety margin of at least 10% over the differential pair saturation voltage. This expression shows that a loop gain value equal to 40 (32 dB) produces a voltage error inferior to 2.5%.

The $V_{BIAS\_N}$ voltage of the replica bias circuit is applied to the $M_{N1}$ transistors of the differential buffers, generating a bias current inside the buffers equal to the bias current of the half-buffer replica circuit, thus the buffers have an output voltage swing with the same value as the $V_o$ voltage[5] of the half-buffer in the replica bias circuit. If the transistors of the differential buffer are scaled to have a larger multiplicity ($M$) than the transistors of the half-buffer, the bias current of the differential buffer will be $M$ times higher than the bias current of the half-buffer. In this case the load resistance value of the buffer is scaled to $R_L/M$ to maintain the buffer output voltage swing equal to $V_{SWING}$.

As in any feedback loop, the replica bias amplifier must be designed to assure that the loop is unconditionally stable. The control voltage $V_{BIASN}$ of the replica bias circuit is essentially a DC signal, since its value is determined by the process corner (that does not change after manufacture) and by the temperature (that changes very slowly). Therefore, in a first approach, the bandwidth of the loop does not need to be large; a bandwidth of a few kHz is enough. However, if the bias current of the buffer needs to be adjusted (to change the buffer delay for example), it is necessary that the replica bias loop bandwidth must be higher than the adjustment frequency. In the analysis in Section 4.2 it was assumed that the delay line control path is modeled by a first order model (expression (4.11)). This means that the closed loop bandwidth of the replica bias circuit for the variable delay buffers of the delay line should be higher than the reference clock frequency of the DLL, if a reference clock frequency of 2 MHz is used, then the replica bias should have a loop bandwidth of at least 8 MHz.

The specifications for the replica bias circuit controlling fixed delay buffers are loop gain larger than 32 dB, gain bandwidth product ($GBW$) larger than 100 kHz and phase margin larger than 60°. The specifications for the replica bias circuit controlling a variable delay are loop gain larger than 32 dB, gain bandwidth product ($GBW$) larger than 8 MHz and phase margin larger than 60°.

The schematic of the replica bias circuit is shown in Fig 4.25. The amplifier is a single stage amplifier because using a low loop gain value (within the limits of the gain specification) improves the phase margin of the loop. The output of the amplifier is obtained mirroring the output current of the amplifier differential pair, this arrangement allows for a low value for the $V_{BIAS\_N}$ voltage.

---

[5] The output voltage of the differential buffer will have a difference to the $V_o$ voltage due to matching errors between the load resistances and transistors of the buffers and the half-buffer. For matched devices in close proximity, matching errors can be as small as .1%, but for distant devices the matching errors are larger and can be up to 5%.

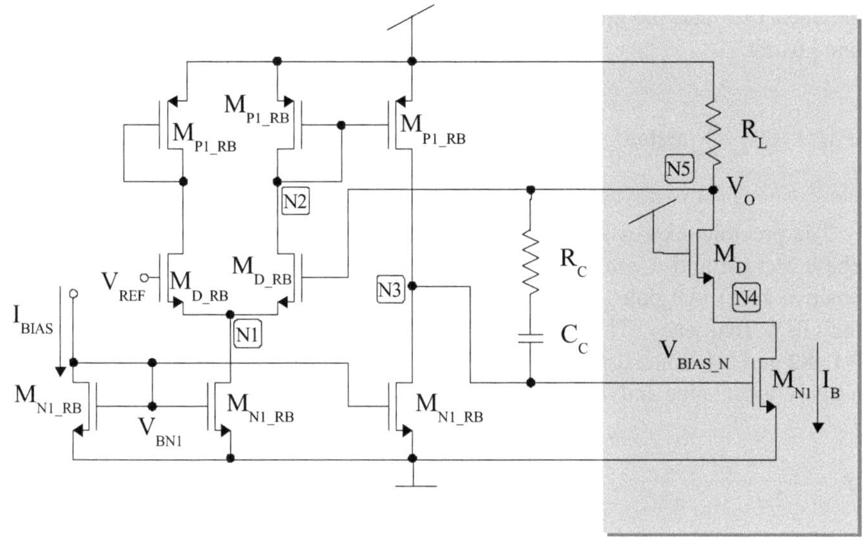


**Fig. 4.25** Replica bias circuit schematic

The loop gain is given by (approximate expression)

$$G_{loop} \approx \frac{gm_{D\ RB} \cdot gm_{N1} \cdot R_L}{g_{ds\ P1\ RB} + g_{ds\ P1\ RB}} \tag{4.41}$$

The loop has a total of 5 poles; which are associated to the 5 nodes of the circuit in the signal path (**N1** to **N5**). The dominant pole of the loop is located, by design, at node **N3,** its approximate expression can be obtained using the Miller theorem and is equal to

$$p_3 \approx -\frac{g_{ds\ P1\ RB} + g_{ds\ N1\ RB}}{C_C \cdot \left(1 + gm_{N1} \cdot R_L\right)} \tag{4.42}$$

The GBW is given by (assuming a phase margin larger than 60°)

$$GBW = p_3 \cdot G_{loop} \approx \frac{gm_{D\ RB}}{C_C} \tag{4.43}$$

The *GBW* of the loop is essentially determined by the value of the compensation capacitor and the value of $g_{m\_N1}$. The phase margin of the loop is determined by all the 5 poles of the circuit and by the two zeros, one is created by the $C_{gs}$ capacitance in the differential pair formed by the $M_{D\_RB}$ transistors and the other is associated to the compensation resistor $R_C$. The phase margin is the difference between 180° and the phase of the loop gain at the *GBW* frequency and is calculated using

$$PM = 180^o - \sum_{i=1}^{5} \arctan\left(\frac{GBW}{-p_i}\right) + \sum_{i=1}^{5} \arctan\left(\frac{GBW}{-z_i}\right) \tag{4.44}$$

The previous expressions shows that a pole contributes with a reduction of the phase margin and a zero with an increase in the phase margin (unless if it is a positive zero). All poles, except $p_3$ are designed to be at a frequency higher than the GBW frequency. This is not very difficult for the poles associated to nodes **N1**, **N2** and **N4**, since these nodes are determined by the ratio between the value of a transconductance and the value of the parasitic capacitance of the node.

$$p_1 \approx -\frac{2 \cdot gm_{D\ RB}}{2 \cdot C_{gs\ D\ RB} + C_{p1}} \tag{4.45}$$

$$p_2 \approx -\frac{gm_{P1\ RB}}{2 \cdot C_{gs\ P1\ RB} + C_{p2}} \tag{4.46}$$

$$p_4 \approx -\frac{gm_D}{C_{gs\ D} + C_{p4}} \tag{4.47}$$

These transconductance values can be by design, in the same order of magnitude as $g_{m\_N1}$ and the parasitic capacitance can be optimized to be much smaller than $C_C$. This is achieved minimizing the values of the channel length of the transistors $M_{N1\_RB}$, $M_{D\_RB}$ and $M_{P1\_RB}$ and maximizing the value $V_{dsat}$ for the same transistors. This strategy results in reducing the value of the parasitic capacitance associated to these transistors and in a loop gain reduction (which is not a problem since the desired loop gain is low). Therefore it is not difficult to obtain frequency values for these poles 5–10 times higher than the desired value of the *GBW*, which results in a phase margin reduction per pole smaller than –11° .It is also important to notice that the pole associated to node **N1** is very close to the zero associated to the $C_{gs}$ capacitance of transistor $M_{D\_RB}$ and therefore they cancel the effects of each other.

$$z_1 = -\frac{gm_{D\ RB}}{C_{gs\ D\ RB}} \tag{4.48}$$

This means that both channel length and $V_{dsat}$ voltage values of the transistors $M_{D\_RB}$ and $M_{N1\_RB}$ are not critical in determining the phase margin value of the loop. The approximated expression of the pole associated to node **N5** is [14]

$$p_5 \approx -\frac{gm_{N1} \cdot C_C}{C_{p3} \cdot C_{p5} + C_{p3} \cdot C_C + C_{p5} \cdot C_C} \quad (4.49)$$

This expression shows, that if the value of $C_C$ is large (3–5 times larger) compared to the values of $C_{p3}$ and $C_{p5}$ then the frequency value of $p_5$ is determined by the ratio of the value of a transconductance and the value of a parasitic capacitance. The value of $C_{p5}$ is small compared to the value of $C_{p3}$, because the $V_{BIASN}$ voltage (node **N3**) is used to bias several buffers and therefore the $C_{p3}$ value is approximately equal to $C_{p3} \approx M \times C_{gs\ N1}$, since the transistors inside the buffers controlled by the replica bias have the same $V_{dsat}$ voltages as the transistors of the half buffer. Under these condition the $p_5$ expression can be further simplified to

$$p_5 \approx -\frac{V_{Dsat\ N1}}{M \times L_{N1}^2} \times K \quad (4.50)$$

where $K$ a is technology dependent parameter. The frequency of $p_5$ does not depend on the $I_B$ value, but on the $V_{dsat}$ voltage of transistor $M_{N1}$ and the $M$ values. This expression shows that the $M_{N1}$ transistor should be designed with a large $V_{dsat}$ and a small channel length (within the design compromises of the differential buffer, described before). The $M$ factor depends on the ratio of the total current in all the buffers controlled by the replica bias and the bias current in the half buffer. Therefore, increasing the number of controlled buffers or the bias current in those buffers, while maintaining the same bias current in the half-buffer, causes a reduction in the frequency of pole $p_5$. This pole, is the second pole of the circuit with the lowest frequency (after pole $p_3$) and therefore it is the pole with the second largest negative contribution to the phase margin (from expression 4.44), resulting that the maximum $GBW$ is located approximately at a frequency 3 times smaller than the $p_5$ frequency. The efficiency of using a replica bias scheme to bias several buffers is related to the desired $GWB$ of the replica bias. If a large $GBW$ is desired, the ratio between the total current of the buffers and the replica bias current must be low to increase the frequency of pole $p_5$ and efficiency is low, if a small $GBW$ is acceptable, the ratio can be high and the efficiency is high.
The value of the compensation zero is given by

$$z = \frac{V_{Dsat\ N1}}{C_C \cdot \left( \frac{1}{gm_{N1}} - R_C \right)} \quad (4.51)$$

The compensation resistance is necessary to transform the positive zero created by the compensation capacitor into a negative zero. The value of $R_C$ is chosen to be $R_C > 1/gm_{N1}$ to obtain a negative zero, the exact value of $R_C$ is determined by simulation and is chosen to compensate part of the phase margin degradation caused by pole $p_5$ and thus improve the maximum *GBW* frequency for a given current ratio, therefore improving the efficiency of the replica bias scheme. The value of the $I_{BIAS}$ current is determined by the required *GBW* of the loop. This current depends on the value of $C_C$, which is determined by the value of $C_{p3}$. Since the $V_{dsat}$ voltage of the transistor $M_{D\_RB}$ can be made smaller than the $V_{dsat}$ voltage of transistor $M_{N1}$, results that $I_{BIAS} \approx I_B$.

Given the required specifications (*GBW*, loop gain and number of buffers to control), the design process for the replica bias circuits is as follows (the same design is used in the two replica bias circuits, one required for controlling the differential buffers and differential gates with fixed current and the second required for controlling the variable delay buffers)

- The transistors inside the half-buffer have the same $V_{dsat}$ voltages and channel lengths as the transistors inside the buffers that the circuit controls. The resistance value is calculated to produce the desired voltage swing at the nominal bias current value.
- The bias current of the half-buffer is selected to obtain a frequency value for the pole $p_5$ 3 times higher than the desired *GBW*.
- The channel length of the transistors $M_{D\_RB}$ is selected to be 0.25 µm to reduce the parasitic capacitance. The $g_{ds}$ of these transistors has a small influence on the gain.
- The $V_{dsat}$ of transistor $M_{D\_RB}$ is selected to be 115 mV to obtain the largest possible gain and *GWB* values for a given current value.
- The channel length of transistor $M_{N1\_RB}$ is selected to be 2 µm to obtain a high $g_{ds}$ value and improve the gain.
- The $V_{dsat}$ of transistor $M_{N1\_RB}$ is selected to be 300 mV to reduce the parasitic capacitance.
- The channel length of transistor $M_{P1\_RB}$ is selected to be 1 µm as a compromise between the $g_{ds}$ value and the parasitic capacitance value.
- The $V_{dsat}$ of transistor $M_{P1\_RB}$ is selected to be 400 mV to reduce the parasitic capacitance therefore maximizing the $p_2$ pole frequency.
- The value of $C_C$ is selected to be 3–5 times larger than the value of the parasitic capacitance in node **N3.** The final value is optimized together with the $R_C$ value using AC simulations. These values are adjusted differently for the two replica bias circuits resulting in $C_C$=8 pF and $R_C$=8 kΩ for the for the replica bias circuit controlling fixed delay buffers and in $C_C$=10 pF and $R_C$=15 kΩ for the for the replica bias circuit controlling a variable delay.

### *4.3.3 Voltage Controlled Delay Line*

The voltage controlled delay line must be able to produce variable delay with a maximum value of 100 ns. Expression (4.33) show that a differential buffer can indeed produce a variable delay, however the method used to vary the delay, changing the load capacitance value, is not easy to implement in CMOS technology. It is easier to change the load resistance of the buffer using MOS transistors in the linear region. The effect on the buffer time constant ($\tau$) will be the same as varying the load capacitance. Varying the resistance will also change the voltage swing of the buffer, however if a replica bias circuit is used the bias current is adjusted to compensate the load resistance variation and the voltage swing will remain constant, this arrangement is shown in Fig. 4.26. The buffer delay is controlled by the $V_{BIASP}$ voltage.

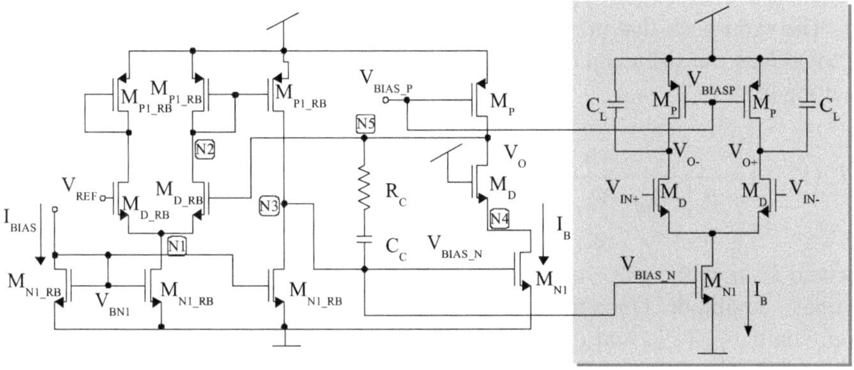


**Fig. 4.26** Replica bias circuit with a variable delay buffer schematic

A PMOS transistor is in the linear region when the $V_{SD}$ voltage is smaller than ($V_{SG}+V_{TP}$),[6] when this condition applies the drain current is given by

$$I_D = \beta \cdot \left[ \left( V_{SG} + V_T \right) \cdot V_{SD} - \frac{V_{SD}^2}{2} \right] \tag{4.52}$$

where $\beta = K_p \cdot W/L$ and $K_p$ is the transistor transconductance which depends on the technology and varies with process and temperature. The drain resistance of the transistor is calculated using

$$r_{sd} = \left( \frac{dI_D}{dV_{SD}} \right)^{-1} = \frac{1}{\beta \cdot \left[ \left( V_{SG} + V_T \right) - V_{SD} \right]} \tag{4.53}$$

From this expression it is clear that a PMOS transistor can be used as variable resistor, whose value is controlled by the $V_{SG}$ voltage. The expression also shows that the resulting resistor has a non-linear dependence on the $V_{SD}$ voltage. The

---

[6] The PMOS transistor voltages are defined with a reverse direction from the voltages of the NMOS transistor to simplify the notation and obtain formulas similar to the NMOS transistor. The threshold voltage ($V_T$) is of course negative.

($V_{SG}+V_T$) voltage should be as large as possible, compared to the maximum $V_{SD}$ voltage, to reduce the non-linear behavior and to guaranty that the transistor is in the linear region. If the PMOS is used to replace the load resistance ($R_L$) in the replica bias circuit of Fig. 4.24, the transistor drain current ($I_B$) will be adjusted until the $V_{SD}$ voltage is equal to the $V_{ref}$ voltage. The drain current of the transistor depends on the $V_{SG}$ voltage value, increasing this voltage value results in increasing the transistor drain current. The maximum value of the drain current (for the desired $V_{ref}$ voltage value) occurs for the maximum value of the $V_{SG}$ voltage, which is of course equal to $V_{DD}$.

The expression that predicts the variable delay of a differential buffer using a PMOS load transistor can be obtained combining expressions (4.33) and (4.53) resulting in

$$T_D\left(V_{biasP}\right)=\frac{0.7\times C}{\beta\cdot\left[\left(V_{DD}-V_{biasP}+V_T\right)-V_{SWING}\right]} \tag{4.54}$$

where $V_{biasP}$ is the gate voltage of the PMOS transistor and $V_{SWING}$ is the output voltage amplitude. Once the output voltage swing is selected, the delay depends only on the $\beta$, $V_{biasP}$ and $C$ values, of these three, only $V_{biasP}$ can be used to vary the delay. The minimum delay is obtained setting the $V_{biasP}$ voltage to 0 V ($V_{SG}=V_{DD}$) and the maximum delay is obtained setting $V_{biasP}=V_{DD}+V_T-V_{SWING}$ ($V_{SG}=V_{SWING}-V_T$) corresponding to the onset of saturation region in the PMOS transistor. If a larger value of $V_{biasP}$ is used, the transistor will enter the saturation region and expression (4.53) will no longer apply, the drain resistance of the transistor in the saturation region does not depend (in first order) on the value of the gate voltage. The ratio between the maximum and minimum delay values depends only on the ratio between the maximum and minimum values of the $V_{biasP}$ voltage, therefore the delay adjustment range is determined by the $V_{biasP}$ voltage range and the maximum (or minimum) achievable delay is determined by the $\beta$ and $C$ values (there is one degree of freedom). The minimum value of $\beta=K_p\cdot{}^{W}\!/_{L}$ is determined by the minimum $W/L$ ratio and the minimum $C$ value is determined by the minimum parasitic capacitance value. Considering that a robust design requires that the capacitance value should be determined by an explicit capacitor instead of a parasitic capacitance (whose exact value is difficult to predict and is layout dependent), the $C$ value should be much larger than the expected value of the parasitic capacitance. Therefore, the C value is set between 300 fF and 0.5 fF (the exact value is adjusted by simulating the circuit and measuring the delay) and the value of the $W/L$ ratio is chosen to obtain the desired maximum delay and a reasonable buffer bias current (see expression 4.52) for the minimum delay case.

The voltage controlled delay line is built using a variable delay buffer followed by a fast buffer to recover the rise and fall times. If only one slow buffer is used to create the variable delay, the problem illustrated in Fig. 4.27 occurs. In this case the desired delay ($T_D$) is 25% of the clock period. The buffers detect a clock transition when the input signal crosses zero. Therefore, for a 25% delay, the $V_{o1}$ voltage must cross zero where is shown in the figure. The output voltage of the slow buffer ($V_{o1}$) is approximated by a ramp in this figure, for simplicity, when in reality it is a damped exponential function (see Section 4.3). In the figure, it is clear that the $V_{o1}$ amplitude is barely equal to the desired amplitude; in reality the amplitude would be inferior to the desired value.

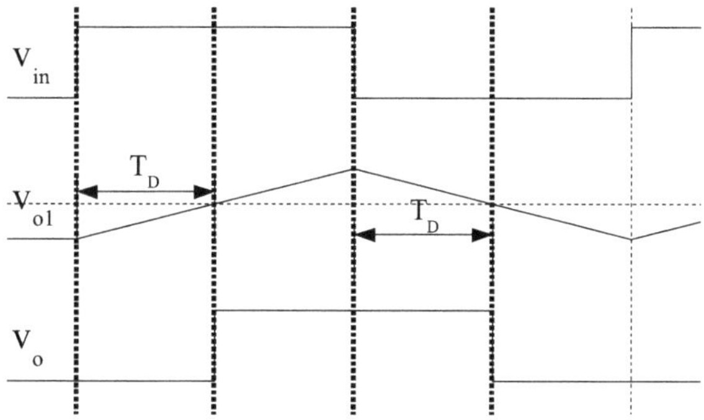


**Fig. 4.27** Signal waveforms in a variable delay line composed by a single slow buffer ($V_{o1}$) and a fast output buffer ($V_o$). The slow buffer output voltage ($V_{o1}$) might not saturate at the digital levels and the fast buffer output ($V_o$) duty cycle might not be 50%

The solution to this problem is to build the variable delay using a cascade of three slow buffers; each buffer produces a delay $T_{Di}$. In this case the delay line can produce a delay ($T_D$) equal to 25% of the clock period and each of the three slow buffers has large saturation time. The signal waveforms for this solution are illustrated in Fig. 4.28.

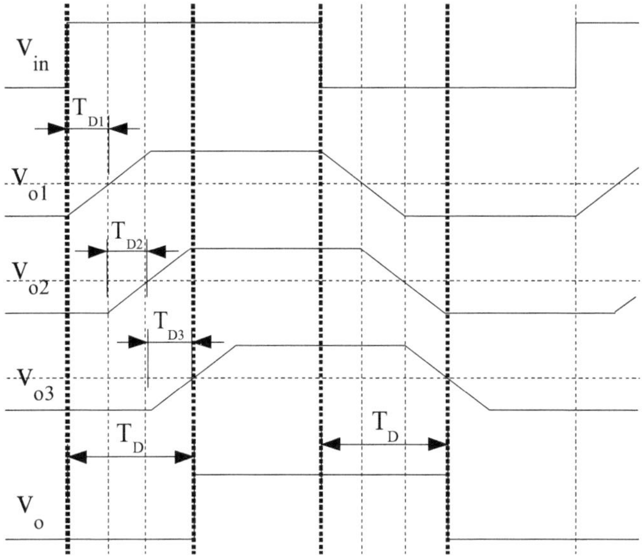


**Fig. 4.28** Signal waveforms in a variable delay line composed by three slow buffers ($V_{o1}$, $V_{o2}$ and $V_{o3}$) and a fast output buffer ($V_o$)

The delay of this delay line is 3 times the delay of a single variable delay buffer plus the delay of the fast buffer at the output. The gain factor of the voltage controlled delay line ($K_{VCDL}$) is given by the derivative of the delay to the control voltage ($V_{biasP}$) and it is proportional to

$$K_{VCDL} = \left( \frac{dT_D}{dV_{biasP}} \right) \propto \frac{1}{V_{biasP}^2} \tag{4.55}$$

The circuit of the delay line is constituted by a cascade of three differential buffers with PMOS loads controlled by a replica bias circuit with a PMOS load. The delay is determined by the value of the $V_{biasP}$ voltage applied both to the replica bias circuit and to the buffer. These three buffers are followed by three fast buffers with resistive loads, controlled by a separate replica bias circuit (with resistive load). The purpose of these fast buffers is to recover the rise and fall times of the clock signal after the slow buffers. The block diagram of the delay line is shown in Fig. 4.29.

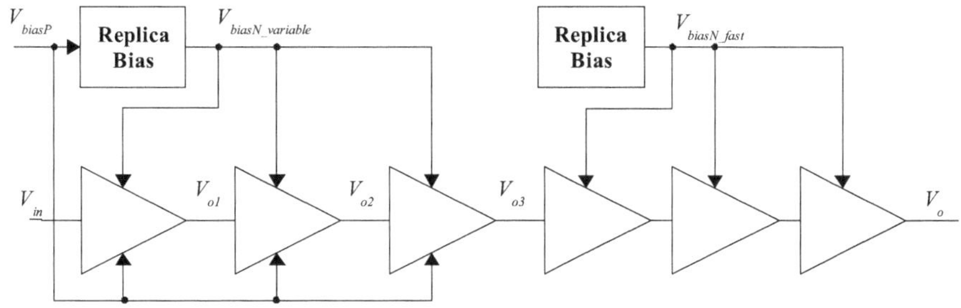


**Fig. 4.29** Block diagram of the Voltage Controlled Delay Line

### *4.3.4 Phase Detector*

The phase detector circuit is a digital state machine with 3 states, whose state transitions are determined by the rising (or falling) edges of the input clock signals [15]. To illustrate the operation of the phase detector, a simplified schematic of a phase detector is shown in Fig. 4.30.

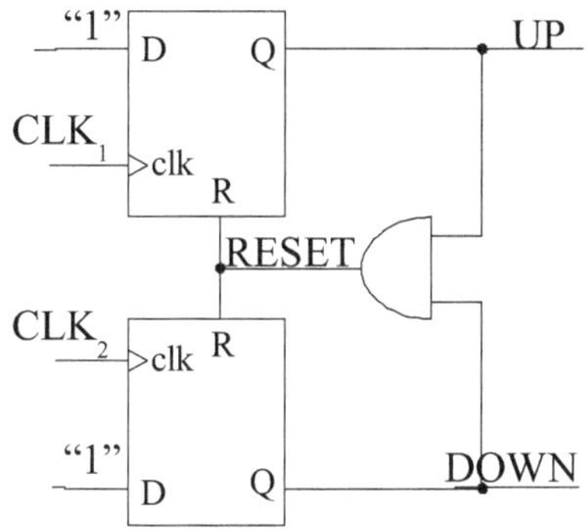


**Fig. 4.30** Simple schematic of a phase detector

The operation of the phase detector is as follows initially the phase detector is in the inactive state (both the *UP* and *DOWN* signals are OFF), when the rising edges of signal $CLK_1$ occurs the flip-flop loads "1" and its output the, *UP* signal, is activated and remains ON until a rising edge occurs in signal $CLK_2$. After the rising edge of signal $CLK_2$ the *DOWN* signal is activated, resulting that both the *UP* and *DOWN* signals are ON, activating the *RESET* signal and the outputs of both flip-flops return to the OFF state. The time interval when the *UP* signal is ON measures the phase difference between the two input clock signals, in this case $CLK_1$ is considered "ahead" of $CLK_2$. If $CLK_1$ is "behind" $CLK_2$, the rising edge of signal $CLK_2$ would occur first, resulting that the *DOWN* signal would be ON instead of the *UP* signal. The maximum amount of delay between the two clock signals with the same period ($T_{CLK}$) is $\pm T_{CLK}$ . In Fig. 4.31 the phase detector signal waveforms are shown for three different delays between the input clock signals, as well as the phase detector characteristic. The phase detector characteristic shows the average activation time (the difference between the *UP* ON time and the *DOWN* ON time) as a function of the delay between the input clock signals. From this characteristic it is clear that the phase detector gain ($K_{PD}$) is equal to one, i.e., a delay $\Delta T$ between the input clock signals results in an activation time equal to $\Delta T$.

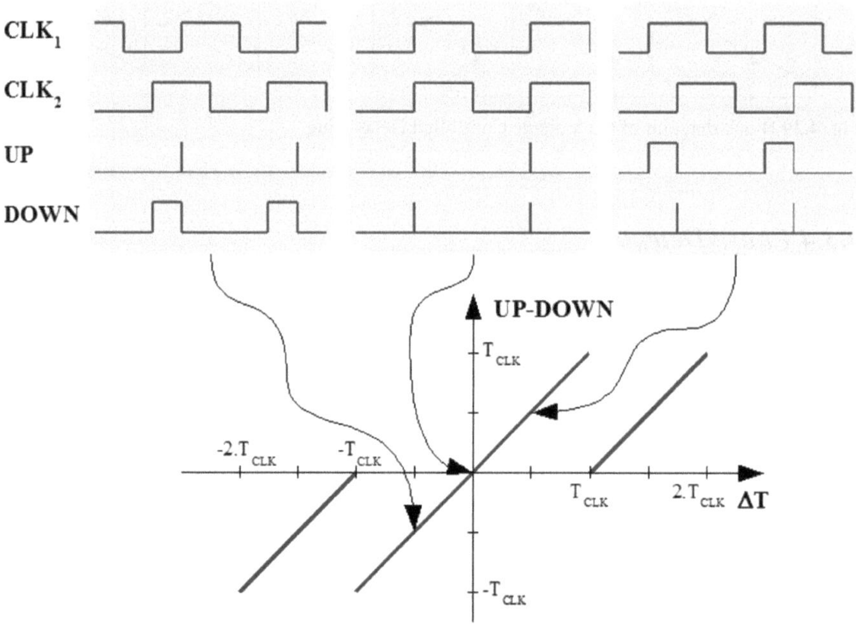


**Fig. 4.31** Phase detector average transfer function and typical operating waveforms

In this circuit, the minimum phase difference detected, is limited by the speed of the flip-flops and the AND gate. If the rising edge of the clock signals is very close, the RESET signal starts to activate before the outputs of the flip-flops have completely settled at the ON value, resulting that the outputs of the flip-flops resets before stabilizing. For very small phase differences this process causes the UP and DOWN signals not to activate at all and the phase difference is not detected. The solution to this problem is very simple [16] if a small delay is added to the reset path, the UP and DOWN signal are always activated (even for zero phase difference) during a minimum time equal to the added delay. Now even a very small phase difference between the input signals is capable of generating a difference between the ON time of the *UP* and *DOWN* signals.

The *UP* and *DOWN* signals are used to control the charge pump that produces the voltage that controls the delay in the delayed clock. Usually, if the *UP* signal is active the amount of delay in the delayed clock increases and if the *DOWN* signal is active the amount of delay decreases.

There is a problem in using a phase detector in a DLL loop [17] the phase detector is a sequential state machine that can start in any of the 3 states. This causes that the first rising edge that occurs (usually that of the reference clock) can be interpreted as the first rising edge (resulting in activating the *UP* or the *DOWN* signal) or as the second rising edge (resulting in resetting the flip-flops returning to phase comparator to the inactive state). Therefore, depending on which state the phase detector initializes, it can see the reference clock as "ahead" or "behind" the delayed clock. This means that the same delay between the two input clock signals can result in either a positive activation time or a negative activation time, in other words the loop gain of the DLL can be positive or negative depending on how the phase detector "wakes up". This problem does not occur in PLL loops, because the period of the loop oscillator clock would become smaller (or larger) than the period of the reference clock and eventually the clock that is "behind" will change place with the other clock and the PLL loop will converge. In a DLL loop, the periods of the reference and delayed clocks are constant and therefore the delay saturates either to the minimum value or to the maximum value. This is a serious problem with a simple solution reset the phase detector to the inactive state at start-up. This is achieved simply by adding an OR gate after the AND gate and applying the new reset signal to that OR gate. It is also important to guaranty that the first rising edge of the reference clock happens before the first rising edge of the delayed clock. This is not difficult since the reference clock delay is either zero or 100 ns ($T_D$) and the delayed clock is between 0 ns and 100 ns.

The phase detector circuit is implemented using differential logic and all the signals are balanced differential signals. The flip-flops are implemented using the circuit shown in Fig. 4.32 [16]

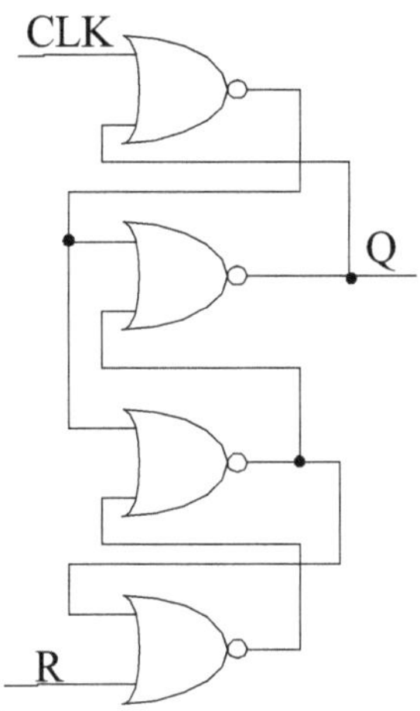


**Fig. 4.32** Flip-flop implementation

The differential gates are implemented using the circuit shown in Fig. 4.33. This circuit is modification of the differential buffer, where a second differential pair is added to obtain a logic function. The circuit can realise the four basic logic operations AND, NAND, OR and NOR, depending on how the input signals are connected, because realizing a NOT function using differential signals consists simply in reversing the polarity of the signal. As an example, the circuit shown in Fig. 4.33 is connected as a AND gate; reversing the polarity of the output nodes, would change the circuit into a NAND gate; reversing the polarity of both input signals, would change the circuit into an OR gate (from the Morgan law) and finally reversing the polarity of the output and input signals, would change the circuit into a NOR gate.

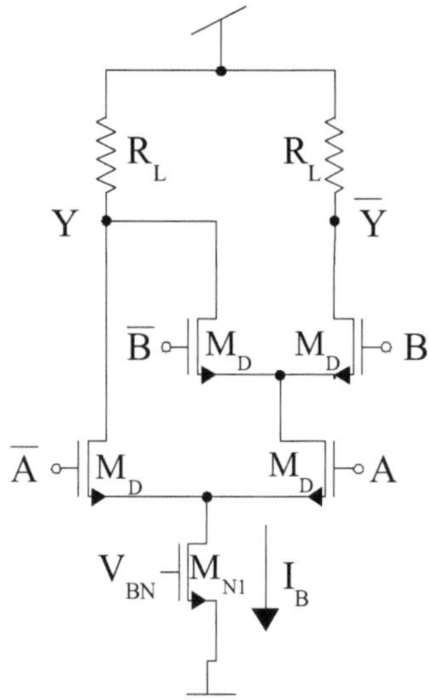


**Fig. 4.33** Differential AND gate circuit

The logic circuits of the phase detector are biased using a common replica bias circuit to all the digital circuits in the DLL. The nominal bias current value of each gate inside the flip-flops is 100 μA and the nominal bias current value of the gates in the reset path is 200 μA.

### *4.3.5 Charge Pump Architecture and Design*

In a single ended charge pump architecture (such as the one depicted in Fig. 4.7) there are two current sources, one sinking current (a NMOS transistor) and another supplying current (a PMOS transistor). The *UP* control signal is active low (the switch is a PMOS transistor) and the *DOWN* control signal is active high (the switch is a NMOS transistor). This asymmetry between the two current sources makes the matching of the turn ON time of the current difficult, it is necessary to add an inverter to one of the control signals causing an extra delay in that control path, the behavior of the NMOS and PMOS switches is different and the value of the up current is not equal to the value of the down current due to matching errors between the current sources.

The selected architecture for the charge pump is completely differential [18], which has several advantages the use of differential digital control signals (*UP* and *DOWN*) eliminates the need to add an inverter to one of the control signals (needed to control a PMOS switch) and therefore the inverted signals are in phase with non-inverted signals; the output current has symmetrical values for the up and down current and the output voltage is more immune to common mode noise. The differential charge-pump architecture is shown in Fig. 4.34

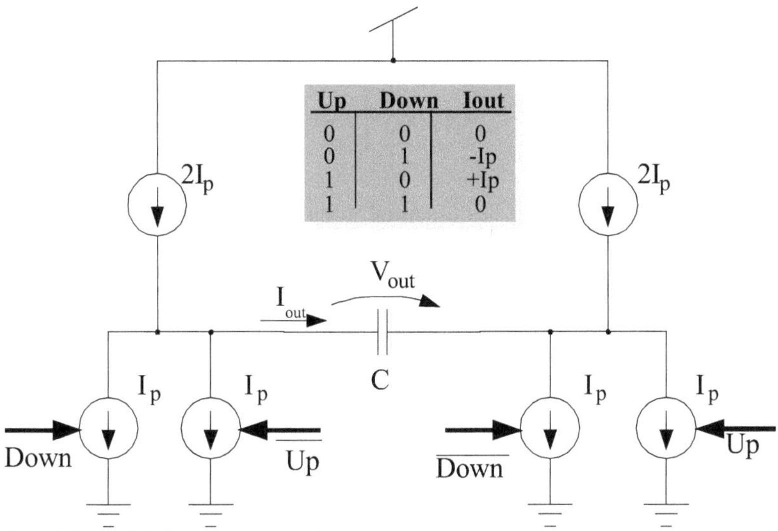


**Fig. 4.34** Differential charge pump architecture

The main disadvantage of the differential architecture when compared to the single end architecture is the lower output resistance of the charge-pump. In the single ended circuit, it is possible to disconnect the current sources from the output node and thus obtain an almost infinite output resistance when the charge-pump current is OFF. In this differential charge pump circuit there are always current sources connected to the output node, which results in a constant output resistance connected to the output node. Considering a finite charge pump output resistance ($R_{out}$), in expressions (4.5) and (4.6) of Section 4.2 results in

$$\begin{aligned} V_{out}\left(n\cdot T_{clk}\right) &= \\ &= V_{out}\left(\left(n-1\right)\cdot T_{clk}\right)\cdot e^{-T_{clk}/(R_{out}\cdot C_1)} + I_p\cdot R_{out}\cdot\left(1-e^{-d[n]/(R_{out}\cdot C_1)}\right)\cdot e^{-(T_{clk}-d[n])/(R_{out}\cdot C_1)} \end{aligned} \quad (4.56)$$

Assuming that ${T_{clk}}/{R_{out}\cdot C_1} \ll 1$ is valid, the previous expression can be simplified to

$$V_{out}[n] \approx \frac{I_P}{C_1} \cdot d[n] + \left(1 - \frac{T_{clk}}{R_{out} \cdot C_1}\right) \cdot V_{out}[n-1] \tag{4.57}$$

This expression is now used to calculate the closed DLL transfer function with the effect of the finite charge pump output resistance. The new closed loop DLL transfer function is given by

$$H_{CL}(z) = \frac{K_{PD} \cdot K_{VCDL} \cdot G \cdot I_P \cdot T_{clk} \cdot z}{K_{PD} \cdot K_{VCDL} \cdot G \cdot I_P \cdot T_{clk} + C_1 \cdot \tau_{RC} \cdot \left(z - 1 + \frac{T_{clk}}{R_{out} \cdot C_1}\right) \cdot \left(z - 1 + \frac{T_{clk}}{\tau_{RC}}\right)} \tag{4.58}$$

This new transfer function has essentially the same frequency response as the one described in (4.13), the main difference is that now the DC gain (obtained making $z=1$) is less than 1. The new DC gain is given by

$$H_{CL}(z=1) = \frac{(K_{PD} \cdot K_{VCDL} \cdot G) \cdot (I_P \cdot R_{out})}{(I_P \cdot R_{out}) \cdot (K_{PD} \cdot K_{VCDL} \cdot G) + T_{clk}} \tag{4.59}$$

The effect of the finite output resistance is that the digital programmable delay will have a gain error in its digital to time conversion characteristic. This gain error is given by

$$error = 1 - \frac{(K_{PD} \cdot K_{VCDL} \cdot G) \cdot (I_P \cdot R_{out})}{(I_P \cdot R_{out}) \cdot (K_{PD} \cdot K_{VCDL} \cdot G) + T_{clk}} = \frac{T_{clk}}{(I_P \cdot R_{out}) \cdot (K_{PD} \cdot K_{VCDL} \cdot G) + T_{clk}} \tag{4.60}$$

The previous expression shows that the gain error can be reduced by increasing the output resistance of the charge pump. Since the drain resistance of a MOS transistor in the saturation region is inversely proportional to the drain current, reducing the charge pump current would accomplish an increase in the output resistance. But as expression 4.60 shows, reducing the charge pump current would cancel any improvements in the gain error obtained from the output resistance increase. In reality it is the factor $I_P \cdot R_{out}$ that should be maximized through a careful design of the current source circuits inside the charge pump.

The gain error can also be reduced by increasing the value of $G$ (gain of the loop transconductance circuit). Increasing this gain also results in reducing the output voltage swing of the charge pump output voltage.

The current source circuits of the charge pump use a folded cascode topology to increase the output resistance and maintain a large output voltage swing for a low power supply voltage. This circuit is depicted in Fig 4.35

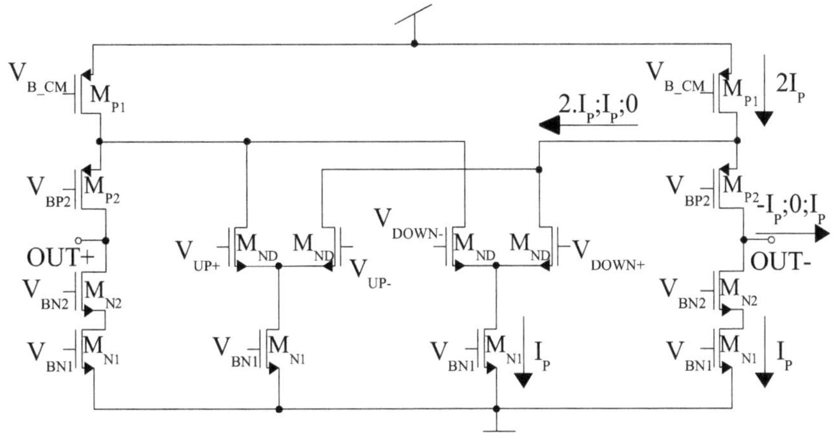


**Fig. 4.35** Folded cascode charge pump circuit schematic

The circuit is an almost direct implementation of the architecture shown in Fig. 4.34. The transistors $M_{P1}$ produce a current equal to $2.I_p$ and the transistors $M_{N1}$ produce a current equal to $I_p$. The output current is determined by the two differential pairs that act as switches steering the current produced by the $M_{P1}$ transistors from the output branch. Transistors $M_{P2}$ and $M_{N2}$ act as cascode devices increasing the output resistance of the current source transistors.
The maximum differential output voltage of the charge pump is given by

$$V_{swing\ CP} = V_{DD} - V_{Dsat\ P1} - V_{Dsat\ P2} - V_{Dsat\ N2} - V_{Dsat\ N1} \tag{4.61}$$

The practical output voltage swing is smaller than the value indicated by the previous expression because it is necessary to add a safety margin to guaranty that all the transistors are well into the saturation and thus obtain a high output resistance. The circuit of Fig. 4.35 is a differential circuit with a high output resistance. This means that any mismatch between the current values of the NMOS and the PMOS current sources would result in the common mode output voltage saturating in either $V_{DD}$ or *GND*. To avoid this situation it is necessary to adjust the current level of the PMOS current sources to match that of the NMOS current sources, this is done using the common mode feedback principle depicted in Fig. 4.36.

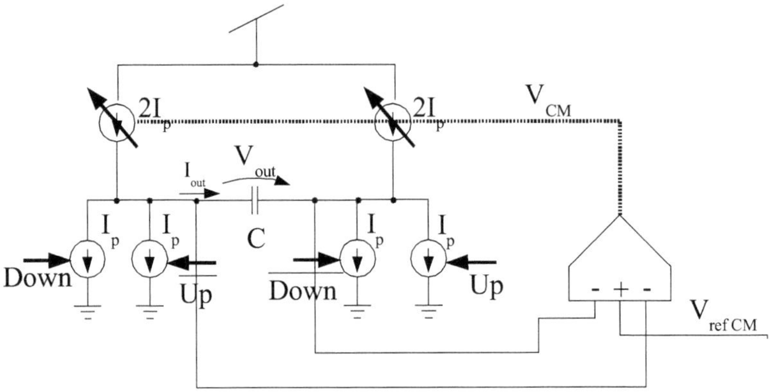


**Fig. 4.36** Common mode feedback circuit scheme

The common mode voltage is adjusted comparing the output common voltage with the desired common mode reference voltage and then using this error voltage to adjust the current of the PMOS current sources, the new current level determines the new output common mode voltage. This operation is equivalent to

$$V_{oCM} = \left( \frac{v_{op} + v_{om}}{2} - V_{CM\,ref} \right) \cdot G = \left( \frac{v_{op} - V_{CM\,ref}}{2} + \frac{v_{om} - V_{CM\,ref}}{2} \right) \cdot G \tag{4.62}$$

where $G$ is the common mode voltage gain of the charge pump. Expression (4.62) suggests that the common mode comparator circuit can be implemented using two differential pairs, each comparing the common mode reference voltage with one of the output voltages. The circuit of the common mode comparator is depicted in Fig. 4.37.

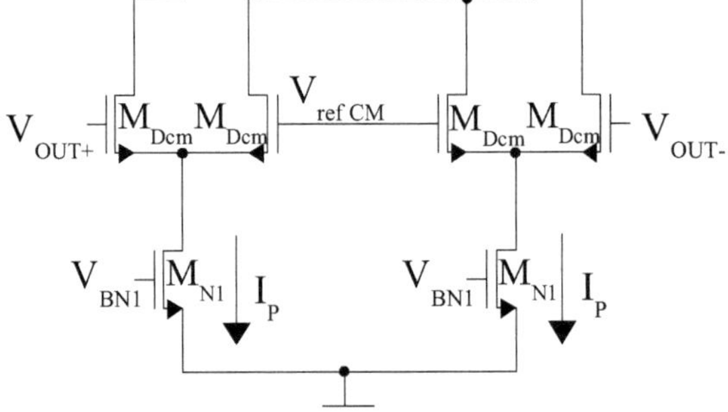


**Fig. 4.37** Common mode comparator circuit

The differential pairs have an output current proportional to the difference between the positive or negative output voltage and the desired common mode voltage level. The maximum output voltage swing of the charge pump is limited by the $V_{dsat}$ voltage of the transistors $M_{Dcm}$. If the output voltage swing is larger than $\sqrt{2} \times V_{Dsat\,Dcm}$ the differential pairs inside the common mode comparator circuit saturate and stops adjusting the output common mode voltage (see Section 4.3.1). The currents produced by the differential pairs are summed in the transistor $M_{P1}$, generating a bias voltage for the transistors $M_{P1}$ inside the charge pump circuit (these transistors should have the same channel length and $V_{dsat}$ voltage).
The minimum value of the desired common mode voltage is limited by the minimum input common mode voltage of the differential pairs of the common mode comparator circuit and is given by

$$V_{CM} \geq V_T + V_{Dsat\,Dcm} + V_{Dsat\,N1} \tag{4.63}$$

The previous expression shows that in order to lower the value of the common mode voltage it is necessary to reduce the output voltage swing of the charge pump. The maximum output voltage swing of the charge pump is set to 300 mV, which results in a minimum output common mode voltage level of 0.9 V.

It is necessary to design both the charge pump circuit and the common mode comparator circuit to guaranty that the common mode feedback loop is stable. The loop must also have a bandwidth larger than the activation frequency of the charge pump (which is the DLL reference frequency) to assure that any common mode voltage wander is corrected within a clock cycle. The common mode small-signal circuit of the charge pump and comparator circuits is shown in Fig. 4.38.

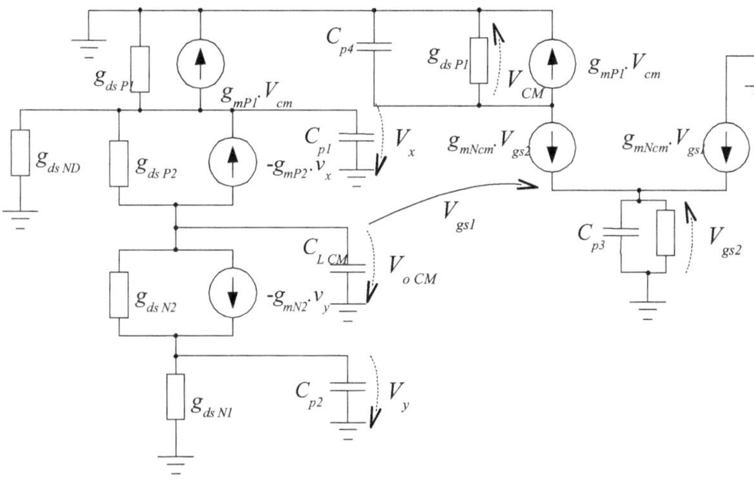


**Fig. 4.38** Common mode small-signal circuit (in closed loop configuration)

The previous circuit is used to calculate the common mode open-loop transfer function; the common mode loop is opened applying an input signal instead of the $V_{oCM}$ voltage.

The output resistance of the charge pump is given by

$$R_{out} = \frac{R_N \cdot R_P}{R_N + R_P}$$
$$R_P \approx \frac{1}{g_{ds\,P1} + g_{ds\,ND}} \cdot \frac{gm_{P2}}{g_{ds\,P2}} \tag{4.64}$$
$$R_N \approx \frac{1}{g_{ds\,N1}} \cdot \frac{gm_{N2}}{g_{ds\,N2}}$$

The common mode loop gain expression can be simplified if it is considered that in the transistors $g_{ds} << g_m$ applies, resulting in

$$G \approx 2 \times gm_{Ncm} \cdot R_{out} \tag{4.65}$$

The common mode loop gain determines the error between the desired common mode voltage and the actual common mode voltage. An error inferior to 1% is acceptable; this corresponds to a gain value larger than 40 dB, which is fairly easy to obtain with this circuit.

The circuit has 4 poles associated to the capacitances $C_{p1}$, $C_{p3}$, $C_{p4}$ and $C_{L\,CM}$. The approximated expressions of these poles are

$$p_1 \approx -\frac{gm_{P2}}{C_{gs\,P2} + C_{db\,P1} + C_{db\,N1}} \tag{4.66}$$

$$p_3 \approx -\frac{2 \times gm_{Ncm}}{C_{gs\,Ncm} + 2 \times C_{s\,Ncm} + C_{db\,N1}} \tag{4.67}$$

$$p_4 \approx -\frac{gm_{P1}}{2 \times C_{gs\,P1}} \tag{4.68}$$

$$p_{out} \approx -\frac{1}{C_{LCM} \cdot R_{out}} \tag{4.69}$$

From the above expressions it is clear that the dominant pole is the pole associated to the output node. Therefore the circuit should be designed to maximize the frequency of all the other poles. Considering that the open loop phase margin is larger than 60° the *GBW* of the loop is determined by the product of the loop gain (expression 4.65) by the output pole (expression 4.69), resulting in

$$GBW \approx \frac{2 \times gm_{Ncm}}{C_{LCM}} \tag{4.70}$$

where $C_{L\ CM}$ is the common mode load capacitance of the charge pump, the value of this capacitance is much smaller than the differential load capacitance and its value is adjusted to obtain the desired *GBW* value.

The charge pump circuit should be designed together with the common mode circuit to achieve the following design goals high output resistance, common mode loop *GBW* larger than 5 MHz (twice the DLL clock), a common mode loop phase margin larger than 60° and a common mode loop gain larger than 40 dB. The design strategy of the charge pump and common mode feedback circuit is as follows

- Transistors $M_{N1}$ and $M_{N2}$ are designed to maximize the $R_N$ value (see expression 4.64), the channel lengths are set to 1.5μm, the $V_{dsat}$ voltage of $M_{N1}$ is set to 200 mV and the $V_{dsat}$ voltage of $M_{N2}$ is set to 120 mV. These transistors can be large since they do not contribute with any poles to the feedback loop.
- The transistor $M_D$ is designed with the same $V_{dsat}$ voltage as the transistors in the differential pairs inside the differential buffers. The channel length is set to 1.5 μm to increase the drain resistance. A large channel length value is possible because the resulting parasitic capacitance of the drain node is still smaller than the parasitic capacitance of the PMOS transistors ($M_{P1}$ and $M_{P2}$) and therefore the parasitic capacitance of this transistor does not dominate in expression (4.66).
- The transistor $M_{P1}$ is designed to have a $V_{dsat}$ voltage equal to 210 mV and a channel length equal 0.8 μm. These values are a compromise between the drain resistance and the parasitic capacitance of the transistor, because this transistor determines the value of two pole frequencies (see expressions (4.66) and (4.68)).
- The transistor $M_{P2}$ is designed to have a $V_{dsat}$ voltage equal to 160 mV and a channel length equal to 1 μm. Once again these values are a compromise to obtain a large output resistance and a small parasitic capacitance. This transistor is very important in determining the output resistance and the pole $p_1$ (see expressions (4.64) and (4.66))
- The bias current of the differential pairs of the common mode voltage comparator circuit is equal to $I_p$, all the transistors inside this circuit except $M_{Ncm}$ have the same sizes as the equivalent transistors inside the charge pump circuit. This reduces the variations between the bias currents and the feedback loop characteristics with process and temperature. The $V_{dsat}$ of transistor $M_{Ncm}$ is chosen to obtain the desired maximum output voltage swing value. The channel length is set to 1.5 μm.

### *4.3.6 DLL Auxiliary Circuits*

The output signal of the charge pump is a differential signal, but the control voltage of the delay line is a single ended signal. Therefore, it is necessary to use a circuit to convert the differential charge pump output voltage to the single ended delay line control voltage. This circuit also implements the second pole in the DLL loop and provides some gain into the loop. The circuit is depicted in the Fig. 4.39.

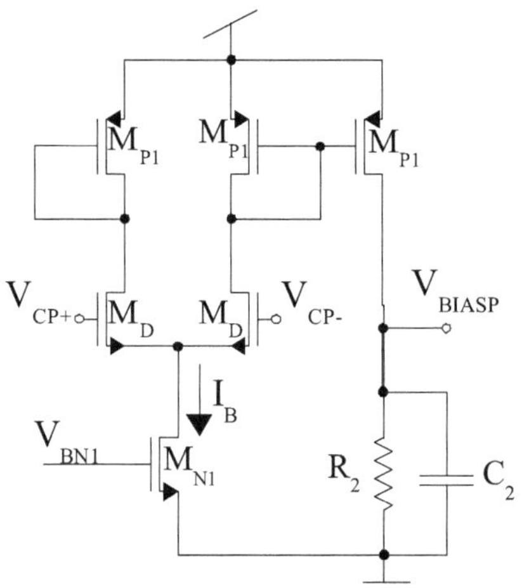


**Fig. 4.39** Differential to single ended conversion circuit and second pole

The circuit is constituted by a differential pair (transconductance) with half of the output current mirrored into a resistor. The output resistor ($R_2$) together with a capacitor ($C_2$) implements the time constant ($\tau_{RC}$) necessary to create the second pole of the DLL loop. The differential pair transistors are designed to have a $V_{dsat}$ voltage large enough to accommodate the output voltage swing of the charge-pump. The circuit has three poles in the signal path the output pole (defined by $R_2$ and $C_2$) ,the PMOS current mirror pole and the pole of the source node of the differential pair. Since the circuit behavior should be characterized by only one pole (expression 4.9), the circuit should be designed so that the frequency of the poles associated to the transistors $M_D$ and $M_{P1}$ is at least 5 times larger than the DLL clock frequency. The bias current of the circuit is determined by the value of the output resistance. This value is selected together with the output capacitance to obtain the desired value of $\tau_{RC}$ and a small area.

The clock signals inside the DLL are differential signals, but the output clock of the DLL needs to be converted to a CMOS clock in order to test experimentally the DLL circuit. The circuit depicted in Fig 4.40 converts the differential clock into a CMOS signal. This circuit is constituted by a two stage amplifier followed by a two CMOS inverters. The amplifier is designed to have a large *GBW*. The first stage is designed to have a large gain to reduce the systematic offset due to the second stage and the CMOS inverters. Since this amplifier works in open-loop there are no stability problems associated to the design.

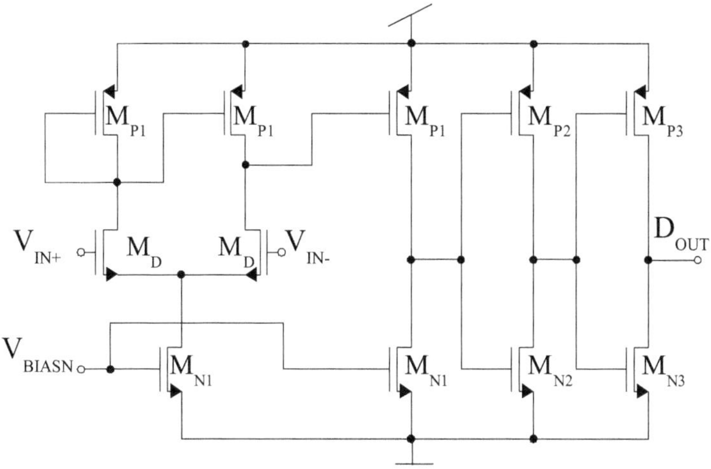


**Fig. 4.40** Differential to CMOS signal conversion circuit schematic

It is also necessary to convert the input CMOS digital signals into differential digital signals. This is done using two differential buffers in series as shown in Fig. 4.41. The input CMOS signal is connected to one input of the first buffer and the other buffer input is connected to a reference voltage ($V_{ref\_dig}$) set to half the voltage swing of the buffer. This differential buffer produces a differential digital signal determined by the input digital signal. The second buffer eliminates any imperfections from the differential signal at the output of the first buffer and improves the rise and fall times of the digital signal. These buffers have a bias current of 100 μA.

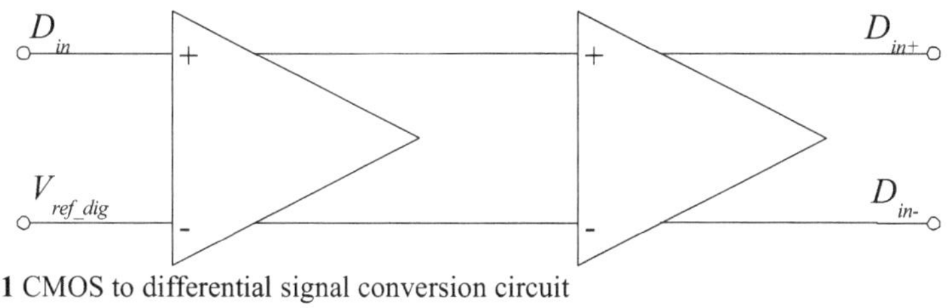


**Fig. 4.41** CMOS to differential signal conversion circuit

## References

1. Mota, M.; Christiansen, J.; A high-resolution time interpolator based on a delay locked loop and an RC delay line; Solid-State Circuits, IEEE Journal of Volume 34, Issue 10, Oct. 1999 Page(s):1360–1366
2. MC10E195, MC100E195 5V ECL Programmable Delay Chip; Data Sheet ON Semiconductor

3. Arkin, B.; Realizing a production ATE custom processor and timing IC containing 400 independent low-power and high-linearity timing verniers; 2004 IEEE International Solid-State Circuits Conference, 2004. Digest of Technical Papers. ISSCC. Volume 1, 15–19 Feb. 2004 Page(s):348–349
4. Design of a digitally programmable delay-locked-loop for a low-cost ultra wide band radar receiver; Paulino, N.; Serrazina, M.; Goes, J.; Steiger-Garcao, A.; Proceedings of the 2003 International Symposium on Circuits and Systems, 2003. ISCAS '03. Volume 1, 25–28 May 2003 Page(s):I-133–I-136
5. Bazes, M., A novel precision MOS synchronous delay line; IEEE Journal of Solid-State Circuits, Volume 20, Issue 6, Dec. 1985 Page(s):1265–1271
6. John G. Maneatis; Low Jitter process independent DLL and PLL based on self-biased techniques, IEEE Journal of Solid-State Circuits, Volume. 31, Page(s): Nov. 1996, 1723–1732
7. van de Beek, R.C.H.; Klumperink, E.A.M.; Vaucher, C.S.; Nauta, B.; On jitter due to delay cell mismatch in DLL-based clock multipliers; 2002. IEEE International Symposium on Circuits and Systems, 2002. ISCAS, Volume 2, 26–29 May 2002 Page(s):II-396–II-399
8. S. Norsworthy, R. Schreier, G. Temes; Delta-Sigma Data Converters, Theory, Design and Simulation; IEEE PRESS; ISBN 0-7803-1045-4
9. A. Bruce Carlson; Communication Systems; McGraw Hill, ISBN 0-07-100560-9
10. Gardner, F.; Charge-Pump Phase-Lock Loops; IEEE Transactions on Communications, Volume 28, Issue 11, Nov. 1980 Page(s):1849–1858
11. N. Metropolis; The Beginning of the Monte Carlo Method; Los Alamos Science Special Issue 1987
12. Chien, G.; Gray, P.R.; A 900-MHz local oscillator using a DLL-based frequency multiplier technique for PCS applications; IEEE Journal of Solid-State Circuits, Volume 35, Issue 12, Dec. 2000 Page(s):1996–1999
13. Yeon-Jae Jung; Seung-Wook Lee; Daeyun Shim; Wonchan Kim; Changhyun Kim; Soo-In Cho; A dual-loop delay-locked loop using multiple voltage-controlled delay lines; IEEE Journal of Solid-State Circuits; Volume 36, Issue 5, May 2001 Page(s):784–791
14. David A. Johns, Ken Martin; Analog Integrated Circuit Design; John Wiley & Sons Inc, ISBN 0-471-14448-7
15. Piguet, C.; von Kaenel, V.; Logic synthesis of a PLL phase frequency detector; IEE Proceedings-Computers and Digital Techniques, Volume 144, Issue 6, Nov. 1997 Page(s):381–385
16. Behzad Razavi; RF Microelectronics; Prentice Hall, ISBN 0-13-887571-5
17. Chulwoo K.; In-Chul H.; Sung-Mo K.; Low-power small-area ±7.28 ps jitter 1 GHz DLL-based clock generator Solid-State Circuits Conference, 2002. Digest of Technical Papers. ISSCC. 2002 IEEE International
18. Rhee, W.; Design of high-performance CMOS charge pumps in phase-locked loops; Proceedings of the 1999 IEEE International Symposium on Circuits and Systems, ISCAS '99., Volume 2, 30 May-2 June 1999 Page(s):545–548

# Chapter 5
# Transceiver Prototype and Experimental Results

**Abstract** this chapter describes the experimental CMOS prototype IC that was developed to evaluate the performance of the different sub-blocks of the radar system. The first part of this chapter is dedicated to the description of the prototype IC and to the design of the short pulse generator circuit and the amplifier circuit needed in the receive path. The second part of the chapter describes the experimental evaluation of the circuit and presents the measured results for each of the radar system sub-blocks. The experimental evaluation procedures used during the testing of the prototype IC, are also described.

In order to evaluate the feasibility of the proposed UWB radar architecture, a prototype integrated circuit containing the most relevant blocks, was designed. This prototype integrated circuit was manufactured in a 0.18 μm CMOS technology with six metal layers and metal to metal capacitors (MiM). The main objective of the prototype is to evaluate the performance of the building blocks of the proposed radar architecture; therefore the prototype integrated circuit needs to have extra pins to access internal signals, which in normal operation would not be necessary. Some sub-blocks of the radar system were not included inside the prototype, in order to reduce the risk of the project and to introduce more flexibility into the experimental evaluation of the circuit. These sub-blocks are the digital logic (necessary to generate the digital clock signals for the digitally programmable delay) and the ADC at the end of the receiver channel. All the sub-blocks were designed to work with a minimum power supply voltage value equal to 1.5 V.

The prototype contains the necessary sub-blocks to build an UWB radar transceiver capable of detecting targets and measuring target ranges up to 15 m. These blocks are: a digital programmable delay, an averaging switched-integrator circuit and a short pulse generator.

## 5.1 Transceiver Circuit

The block diagram of the proposed UWB radar system is depicted in Fig. 5.1.



N. Paulino et al., *Low Power UWB CMOS Radar Sensors*, 167–209.

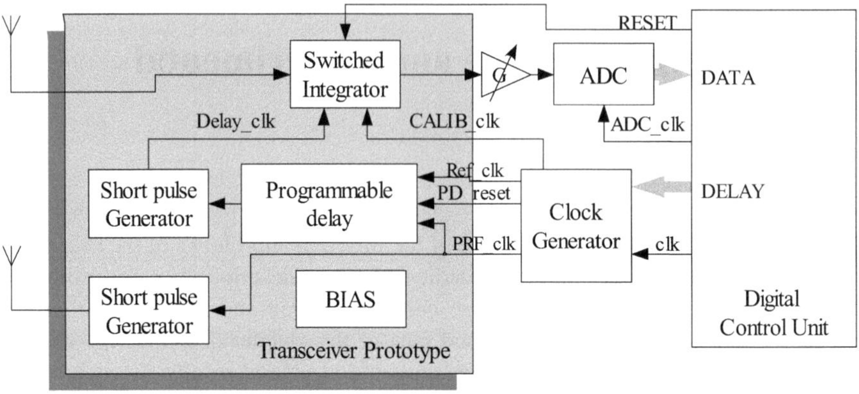


**Fig. 5.1** Block diagram of an UWB radar system. The blocks inside the gray rectangle are integrated into the prototype circuit

The radar system is constituted by a short pulse generator that creates the UWB signal that is radiated, by a switched-integrator circuit that receives the echo signals, by a programmable delay circuit that determines the range of the target and by a digital control unit that controls the operation of the entire radar system. There is also a bias block that contains a replica bias circuit needed to provide the $V_{biasN}$ voltage for all the differential logic circuits. The radar system uses two separate antennas, one for the transmit channel and one for the receive channel, in order to reduce coupling problems. The receive channel is constituted by a switched-integrator circuit with sample averaging, followed by an adjustable gain and an ADC (this gain is used to adjust the switched integrator output voltage range to the ADC input voltage range). The voltage gain of the receive channel can be controlled by the number of samples that are added together in the switched-integrator circuit. The sampling instant is determined by the amount of delay in the sample clock of the switched-integrator (controlled by the programmable delay circuit). This delay value determines the range where the radar will look for a possible target. This parameter is controlled by the digital control unit; this unit determines how the target range is swept. The range can be swept sequentially or if necessary certain ranges can be avoided. The receiver gain can be adjusted to an ideal value for each particular range by changing the number of samples added together (using the RESET signal together with the DELAY and ADC_clk signals). The digital control unit can also realize complex signal processing operations on the received signal. These operations can be realized by a dedicated DSP or by a host computer. In that case, the digital control unit would serve mainly as the interface block, performing only some basic interfacing functions and providing the master clock signal.

All the sub-blocks in the radar system can be manufactured using CMOS technology and can be, in theory, integrated into a single silicon die. However, in order to facilitate the experimental evaluation of the system sub-blocks, only the sub-blocks inside the gray box in Fig. 5.1 are integrated into the CMOS prototype transceiver circuit. This partition of the radar system allows for the complete test of the analog blocks (which is the main objective of the prototype circuit) and also allows to build a complete radar system using the prototype circuit and additional circuits. The clock generator sub-block is a completely digital circuit that generates all the necessary clock signals for the analog blocks. The digital sigma-delta modulator, that creates the reference clock for the programmable delay, is inside this block. This block can easily be implemented using external digital hardware (such as a FPGA) or a digital pattern generator. Therefore it will not be included inside the experimental prototype IC. The ADC converts the analog output voltage of the switched integrator into the digital domain. This block can be implemented using a discrete ADC or instead a digital oscilloscope can be used to simplify the test set-up of the circuit.

The sub-blocks in the transceiver are designed to obtain a radar system with a maximum range of 15 m and a range resolution of 3 cm. The digitally programmable delay circuit has a maximum delay of 100 ns and a 9 bit programming resolution. In order to obtain the required precision in the delay values, it is necessary to create all the clock signals from a clock with a period equal to 200 ns, as shown in Fig. 4.3 of the previous chapter. The clock signals of the radar system have a period equal to 400 ns, corresponding to a frequency value of 2.5 MHz. These include the PRF clock, which controls the time instants when the pulses are radiated, and the sample clock, (the delayed version of the PRF clock) which controls the sampling instants, they have the same frequency value. The receiver channel of the radar system is designed to have a maximum input referred noise voltage smaller than 10 μVrms. The number of averaged samples can be adjusted for each range. If 100 samples are averaged for each range, the entire radar 15 m range can be swept in $400\,ns \times 100 \times 512 = 204.8$ ms, which corresponds to approximately 5 sweeps of the radar range per second.

## 5.1.1 Short Pulse Generator

As shown in Chapter 2, circuits built using CMOS technology are capable of generating short pulses with a duration inferior to 1 ns. A simple circuit for generating short pulses is depicted in Fig. 2.10 of Chapter 2. This circuit is simply an AND gate where one input is connected to a clock signal and the other input is connected to a delayed version of the clock signal. The pulse obtained using this circuit is depicted in Fig. 2.11 and has a duration inferior to 1 ns. The shape of the pulse is determined by the rise and fall times of the AND gate together with the load of the circuit (the antenna). However, the previous circuit has a problem: it creates pulses both on the rising and falling edge of the clock signal. In order to eliminate this problem, it is necessary to use an edge triggered flip-flop, when the rising edge of the clock appears the flip-flop changes its output signal from “0” to “1”. The output is connected to the asynchronous reset of the flip-flop and the result is that the output signal changes from "1" to "0". The end result is a short rectangular pulse produced only in the rising edge of the clock signal. This circuit is shown in Fig. 5.2.

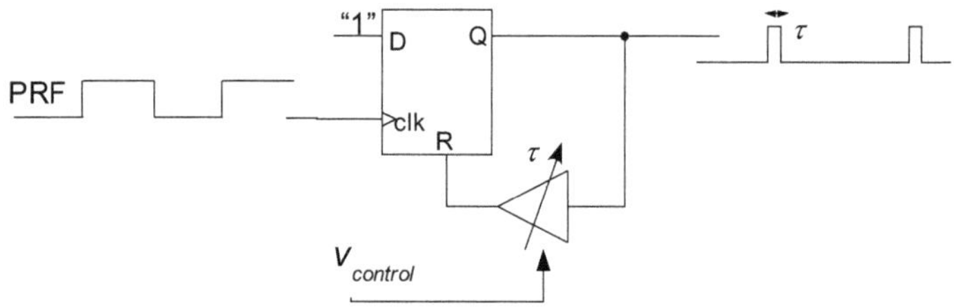


**Fig. 5.2** Simplified block diagram of the short pulse generator circuit

In order to control the duration of the pulse, a variable delay buffer is introduced between the flip-flop output and the flip-flop reset. The amount of delay in this buffer is controlled by a control voltage. Another alternative is to use a digitally programmable delay, such as the one described in the previous chapter, to generate a precise and controlled delay value. The short-pulse generator circuit integrated in the prototype circuit does not include this programmable delay because it would occupy a large area. Since the objective is to test the several sub-blocks, the short-pulse generator can be tested using an external analog voltage to control the pulse duration.

The short pulse generator circuit uses, as much as possible, circuits that are already designed for the other sub-blocks. Therefore the flip-flop is the same as the one used in the charge pump of the DLL, the schematic of this flip-flop is depicted in Fig. 4.32 of the previous chapter. This flip-flop uses 4 differential NOR gates. The schematic of these gates is shown in Fig. 4.33 of the previous chapter. The variable delay buffer is constituted by a single slow inverter designed for the VCDL (see Section 4.3.3 of the previous chapter). The voltage swing value of this delay buffer is adjusted by a dedicated replica bias circuit (with a schematic equal to the one depicted in Fig. 4.26 of the previous chapter) so that it is kept constant for the different delay values. The output signal of the flip-flop goes through two buffers with increasing driving capability, in order to decouple the value of the pulse duration from the load of the circuit. These inverters also improve the rise and fall times of the pulse. The short pulse generator circuit uses differential circuits, which are class A circuits (the bias current is constant). If these circuits were to drive directly a small impedance (such as the antenna) it would be necessary to use a differential buffer with large bias current value in order to obtain small rise and fall times. Since the clock frequency (PRF) is low (2.5 MHz), it is more efficient to use a CMOS buffer because CMOS circuits only draw current during the voltage transitions of the signal. Therefore, in this case, a CMOS buffer uses, on average, much less current than a differential inverter with the same rise and fall times. In order to use a CMOS inverter it is necessary to convert the output differential signal of the flip-flop into a CMOS signal. This task can be accomplished using the differential to CMOS signal conversion circuit depicted in Fig. 4.40 of the previous chapter. The output of this circuit is then connected to two CMOS inverters (with an increasing driving capability) before driving the antenna. The complete short pulse generator schematic is depicted in Fig. 5.3.

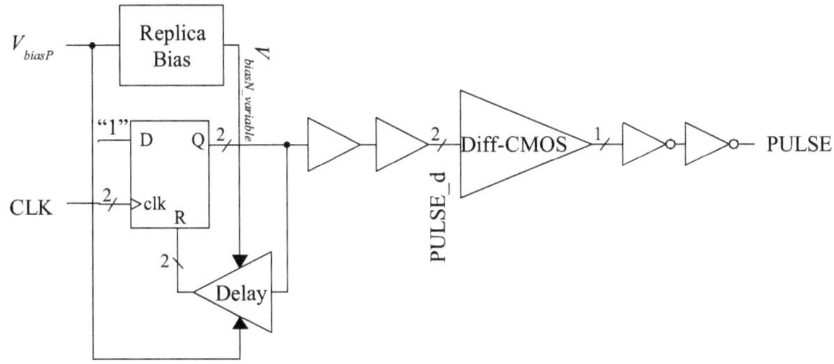


**Fig. 5.3** Schematic of the short pulse generator circuit

The $V_{biasN}$ voltage needed for all the differential gates and buffers (except the variable delay buffer) of this circuit is provided by the shared replica bias circuit of the integrated prototype. The variable delay in the short pulse generator buffer is controlled by a dedicated replica bias circuit such as the one depicted in Fig. 4.26 of the previous chapter. By adjusting the $V_{biasP}$ voltage value in this replica bias circuit, the short pulse generator is capable of generating pulses with a duration from 0.5 ns to 2 ns and an amplitude of 1.5 V, as shown in Fig. 5.4.

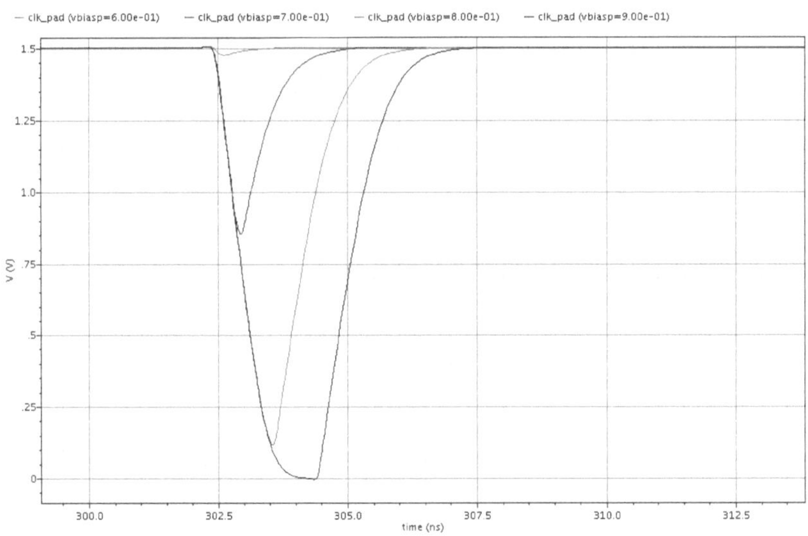


**Fig. 5.4** Simulated output signal of the short-pulse generator circuit for different $V_{biasP}$ voltage values and a load composed by a 13 pF capacitor in parallel with a 1 MΩ resistor

The short pulse generator was designed to drive a 13 pF load and produce a pulse with 1.5 V of amplitude. This amplitude value can be adjusted using external circuits if necessary. If a biconical antenna is used (such as the one described in Section 2.3.3) a pulse with an amplitude value of 0.5 V should be enough to meet the FCC requirements (see Fig. 2.26) and the output pulse must be attenuated. If a different antenna is used, or if the pulse power needs to be dramatically increased (in the case of using the radar transceiver in a GPR application), this pulse signal can be used to drive an external power amplifier.

The short pulse generator circuit is also used to create the differential clock that controls the integrating interval of the switched integrator. In this case, the differential output of the circuit (PULSE_d) is used and therefore it is not necessary to use the differential to CMOS signal conversion circuit and the CMOS inverters. Due to the calibration cycle, the switched integrator needs two differential short pulse signals, one when the calibration is active and the other when the calibration is off (see Section 3.3.9 of Chapter 3). The switched integrator in reality does not require two separate differential clock signals; since the $v_{clkm}$ signal of the differential clock is always connected to the same transistor (see Fig. 3.31 of Chapter 3). Therefore, it is only necessary to create two separate $v_{clkp}$ signals; this is done using an analog multiplexer. The multiplexer produces two clock signals ($v_{clkp1}$ and $v_{clkp2}$) from one $v_{clkp}$ clock signal. The multiplexer circuit is shown in Fig. 5.5. The multiplexer is constituted by two pairs of switches, each pair contains a series switch (transistor $M_{P1}$ or transistor $M_{P2}$) and a switch to ground (transistor $M_{N1}$ or transistor $M_{N2}$) controlled by the logical signal CALIB. It is necessary to use the switch to ground to guaranty that the transistors $M_{s2a}$ or $M_{s2}$ are off when the clock signal is applied to the other transistor (see Fig. 3.31). The signal $v_{clkpm}$ must go through a similar circuit (transistor $M_{N0}$ OFF and transistor $M_{P0}$ ON) to guaranty that it has the same delay and distortion as the $v_{clkp1}$ and $v_{clkp2}$ clock signals and therefore still form a perfect differential signal.

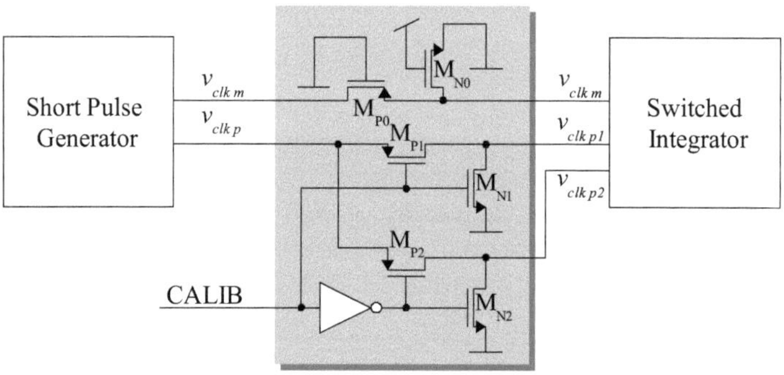


**Fig. 5.5** Schematic of the analog multiplexer used to create the differential clock signals for the switched integrator circuit

A timing diagram with the clock signal waveforms of the analog multiplexer is shown next in Fig. 5.6 and the simulated differential clock signal ($v_{clkp1}$-$v_{clkm}$) for different values of the $V_{biasP}$ voltage is shown in Fig. 5.7.

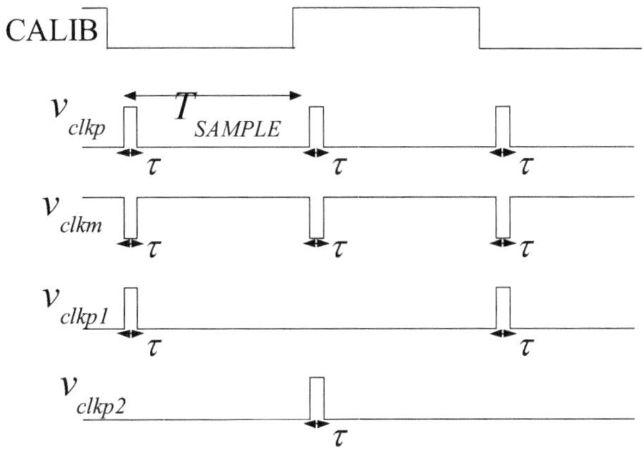


**Fig. 5.6** Timing diagram of the clock signals of the analog multiplexer

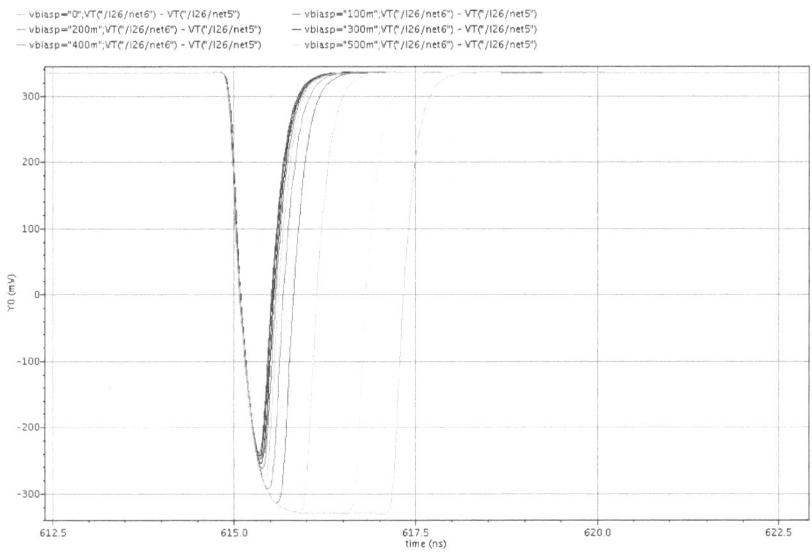


**Fig. 5.7** Simulated differential clock signal for different values of the $V_{biasP}$ voltage

The expected supply currents of the sub-circuits of the short pulse generator (for the typical case) are shown in Table 5.1.

**Table 5.1** Expected supply current of the short pulse generator sub-circuits

| Circuit | Flip-flop | Replica bias | delay | buffer | buffer | Diff-CMOS | Total |
|---|---|---|---|---|---|---|---|
| Current | 400 μA | 400 μA | 100 μA | 400 μA | 800 μA | 200 μA | 2300 μA |

### *5.1.2 Receiver Channel*

The receiver channel is composed by the switched integrator circuit followed by an ADC. As previously discussed (see Section 3.3.6 of Chapter 3), the antenna is directly connected to the input of the switched integrator, instead of being connected first to a LNA. This solution is possible only if the physical distance between the antenna and the circuit is smaller than the transmitted pulse width (approximately 30 cm for a pulse with 1 ns duration), otherwise there would exist multiple reflections of the input pulse due to the impedance mismatch between the antenna and the switched-integrator input. If the antenna is physically distant from the receiver circuit, connected by a long cable (transmission line), it would be necessary to use a LNA close to the input of the switched integrator. This LNA would be designed to have a constant impedance (matched to the cable impedance) over the bandwidth of the transmitted pulse in order to avoid multiple reflections of the input signal. Using the LNA circuit would degrade the input referred noise of the receiver channel and would increase the power dissipation of the radar system.

As shown in Fig. 5.8, the switched integrator circuit is constituted by a differential switch, by a load circuit and by two capacitors ($C_1$). The differential switch produces a differential charge, proportional to the average input signal during the time interval when the switch is closed ($\tau$). This charge is injected into the capacitors $C_1$, producing the differential voltage $v_{out1}$. This voltage is in turn, amplified by the differential amplifier to produce the differential output voltage $v_{out2}$. During this amplification, the charge in capacitors $C_1$ is transferred to capacitors $C_2$, resetting the $v_{out1}$ differential voltage. Finally, this voltage is converted into a binary number with $n$ bits of resolution by the ADC.

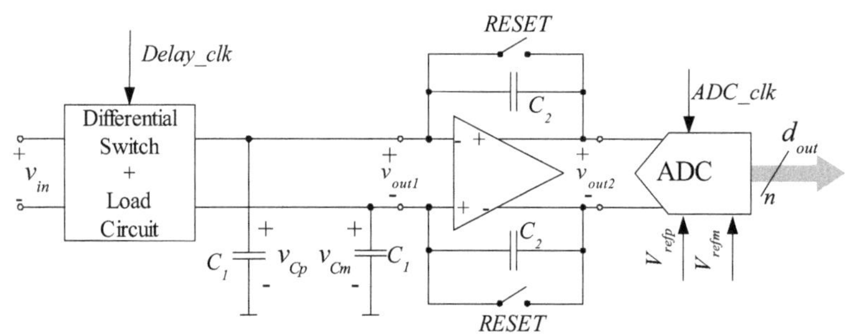


**Fig. 5.8** Simplified schematic of the receiver channel

The signal amplitudes and noise levels in the receiver circuit can be calculated using the signal-flow diagrams shown in Fig. 5.9.

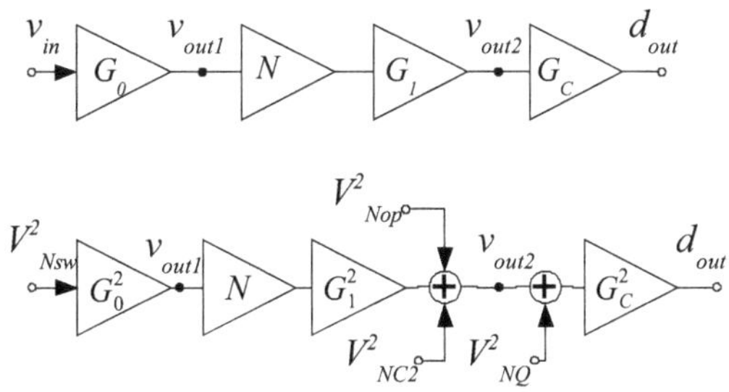


**Fig. 5.9** Signal diagram of the receive channel (top) and Noise signal diagram of the receiver channel (bottom)

The input signal is amplified by the switched integrator circuit ($G_0$ given by expression 3.102 of Chapter 3), by the number of samples added together ($N$), by the amplifier gain ($G_1 \approx C_1/C_2$) and finally by the conversion gain of the ADC ($G_C$) given by:

$$G_C = \frac{d_{out}}{v_{out2}} = \frac{1}{V_{ref}} \cdot 2^{n-1} = \frac{1}{V_{lsb}} \tag{5.1}$$

where $V_{ref}$ and $n$ are the reference voltage (maximum input amplitude) and resolution of the ADC, respectively. The total gain in the signal path is given by:

$$G = G_0 \cdot N \cdot G_1 \cdot G_C \tag{5.2}$$

The value of this gain can be adjusted by changing the pulse width of the clock signal of the switched integrator circuit (thus changing the value of $G_0$) or by changing the number of samples ($N$) averaged together. The value of the gain can be controlled independently for each possible target range, by adjusting the number ($N$) of samples averaged for that particular range (using the RESET signal).

The output noise level of the receiver channel can be calculated using the noise signal diagram depicted on the bottom of Fig. 5.9. There are three main noise sources in this diagram: the input referred noise voltage of the switched-integrator circuit ($V_{Nsw}$ given by expression 3.101 of Chapter 3), the output noise voltage of the amplifier ($V_{Nop}$), the noise created by the reset switch in capacitor $C_2$ ($V_{NC2} = \sqrt{K_B \cdot T/C_2}$) and the quantization noise of the ADC ($V_{NQ} = V_{lsb}/\sqrt{12}$). The total output noise of the receiver channel is given by:

$$V_{Nout} = \sqrt{V_{Nsw}^2 \cdot G_0^2 \cdot N \cdot G_1^2 \cdot G_C^2 + V_{Nop}^2 \cdot G_C^2 + V_{NC2}^2 \cdot G_C^2 + V_{NQ}^2 \cdot G_C^2} \tag{5.3}$$

The input referred noise of the receiver circuit can be obtained dividing expression 5.3 by expression 5.2, resulting in:

$$V_{Nin} = \frac{V_{Nout}}{G} = \sqrt{\frac{V_{Nsw}^2}{N} + \frac{V_{Nop}^2}{G_0^2 \cdot N \cdot G_1^2} + \frac{V_{NC2}^2}{G_0^2 \cdot N \cdot G_1^2} + \frac{V_{NQ}^2}{G_0^2 \cdot N \cdot G_1^2}} \quad (5.4)$$

this expression shows (as expected) that if the value of $N$ is large the input referred noise value is mainly dependent on the input referred noise of the switched integrator circuit. As previously discussed, the radar should be able to detect target located at a 15 m range. Although different targets can result in larger or smaller amplitudes for the returning echo signals, the analysis of possible echo signals in Section 3.2 of Chapter 3 suggests that a reasonable value for the minimum amplitude of the echo signal is around 10 μVrms. This means that the receiver channel should have a maximum input referred noise equal to 10 μVrms (noise budget) in order to be able to detect the echo signal. The available noise budget should be partitioned between the different circuits composing the receiver channel. Most of this noise budget is attributed to the differential switched-integrator circuit, since the input referred noise is dominated by this block.

The switched-integrator circuit was designed in order to have an input referred noise value smaller than 10 μVrms, after 100 output samples are added together. The input referred noise voltage of the switched integrator (without using averaging) is plotted in Fig. 3.33 of Chapter 3, for different values of bias current $I_B$ and $V_{Dsat}$ voltages. The input referred noise voltage of the circuit after 100 samples are averaged can be estimated simply by dividing the input referred noise voltage value from the graph divided by 10. This graph is used to design the switched-integrator circuit. Given the area limitation, the bias current is set to 2 mA, resulting in an estimated input referred noise for the switched-integrator equal to 81.7 μVrms (using a $V_{Dsat}$ voltage equal to 120 mV in transistor $M_1$). As explained in Chapter 3, the value of capacitance $C_1$ must be larger than the limit given by expression 3.80. Allowing for a common mode voltage drop of 100 mV and assuming a $\tau$ value equal to 1 ns, the previous expression returns a minimum value for $C_1$ equal to 5 pF. However, since this circuit is an experimental prototype, where the value of $\tau$ is not necessarily fixed to 1ns, the value of $C_1$ is set to 10pF to allow for a greater variation of both $\tau$ and the bias current $I_B$ during the experimental evaluation of the circuit. The voltage gain value of the switched integrator circuit ($G_0$) can be calculated using expression 3.102 and it is approximately equal to 0.67 (considering $\tau$= 1 ns, $C_1$= 10 pF, $C_D$>>$C_1$ and $R_S.C_{gs}$.>>$\tau$).

The remaining of the noise budget of the receiver channel is attributed to the amplifier circuit and the ADC. The noise level of the ADC is determined by its resolution and reference voltage values. The maximum reference voltage value is limited by the maximum voltage swing in the ADC, which in turn is limited by the power supply voltage value. Considering that the ADC uses an amplifier with only two transistors in the output stage, the maximum peak to peak differential voltage swing (corresponding to $2.V_{ref}$) would be given by an expression similar to expression 3.93 of Chapter 3. Therefore the maximum value of $V_{ref}$ would be around 1 V, for a power supply voltage value of 1.5 V. The ADC resolution is limited by the accuracy of the matching between different components (capacitors or resistors) in the selected CMOS technology. Depending on the technology the maximum ADC resolution can be between 9 and 12 bits (assuming that calibration or sigma delta modulator techniques are not used). Therefore in this analysis it will be assumed that the ADC reference voltage is equal to 1 V and the ADC resolution is equal to 10 bits.

In order to determine the available noise budget for the amplifier, the input referred noise of the receiver channel is calculated (using expression 5.4) for different values of the amplifier output noise and of the amplifier gain. In the next graphs all the switched integrator parameters, except the number of samples added together ($N$), are fixed to: $I_B$=2 mA, $V_{DsatM1}$=120 mV, $V_{DsatM2}$=300 mV $\tau$= 1 ns and $C_1$=10 pF. In Fig. 5.10 the input referred noise of the receiver channel is plotted as a function of the output noise voltage of the amplifier ($V_{Nop}$), the amplifier gain ($G_1$) is equal to 10.

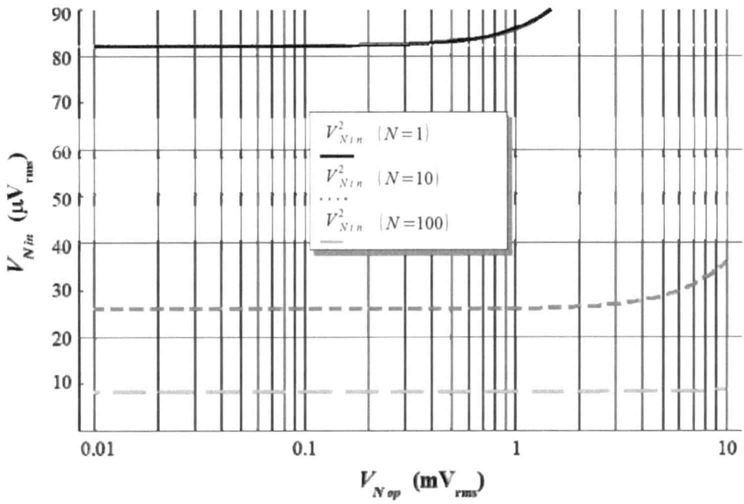


**Fig. 5.10** Receiver channel input referred noise as a function of the amplifier output noise voltage ($G_1$=10), for 3 different values of $N$

The previous graph shows that the input referred noise of the receiver channel can be below 10 µVrms if the 100 samples are added together and the output noise voltage of the amplifier is smaller than 10 mVrms. In Fig. 5.11 the input referred noise of the receiver channel is plotted as a function of the amplifier gain ($G_1$), the output noise voltage of the amplifier ($V_{Nop}$) is equal to 1 mV.

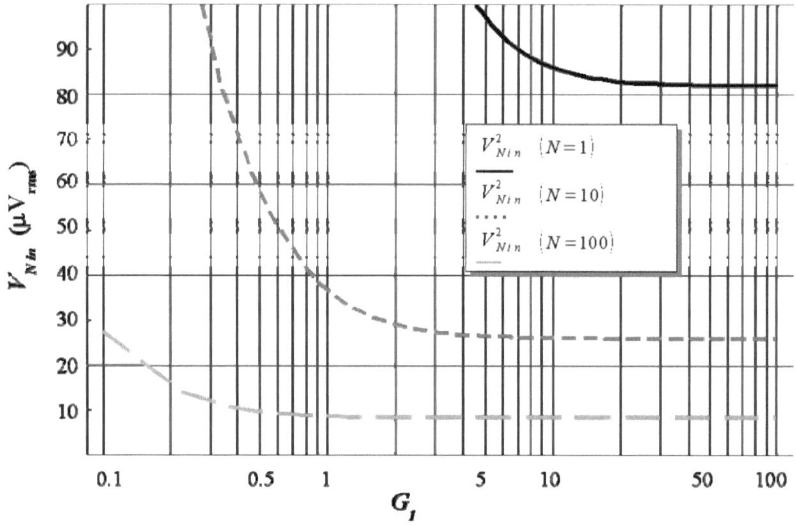


**Fig. 5.11** Receiver channel input referred noise as a function of the amplifier gain ($G_1$) for an amplifier with $V_{nop}$=1 mVrms and for 3 different values of $N$

The previous graph shows that the receiver input referred noise decreases with the increase of the gain $G_1$. After a certain value of $G_1$ the reduction of the input referred noise is marginal, therefore the value of $G_1$ is set to 10. With this gain value and an amplifier with an output noise voltage value equal to 10 mVrms, results that the input referred noise value of the receiver channel is equal to 8.6 mVrms.

During the offset calibration phase, it is necessary to shunt the differential input signal through a switch (transistor $M_{s4}$ in Fig. 3.31 of Chapter 3). This switch must have an ON resistance value significantly smaller than the signal source resistance; therefore its ON resistance should be smaller than 1 Ω. But since the common mode input voltage of the switched integrator circuit is high (it must be larger than $V_T + V_{Dsat\,M1} + V_{Dsat\,MB}$), the $V_{GS}$ voltage of the $M_{s4}$ transistor is small and therefore, the transistor must have a large channel width to obtain the desired low value for the ON resistance. Such a large transistor would introduce large parasitic capacitors in the input nodes, resulting in a significant attenuation of the input signal. In order to avoid using such a large transistor, it is necessary to increase the $V_{GS}$ voltage value. This can be achieved using two transistors to shunt the input signal to ground, in this case the $V_{GS}$ voltage is equal to the power supply voltage ($V_{DD}$). Since this new arrangement effectively grounds the gates of the transistors $M_1$, the switched-integrator circuit would not work properly during the calibration phase. This problem is solved easily by AC coupling the input signal to the gates of the transistors $M_1$ through two capacitors ($C_D$), as shown in the new schematic of the switched-integrator circuit (depicted in Fig. 5.12). The correct bias voltage for the gate of the transistors $M_1$ is provided by two diode connected transistors ($M_{B1}$) each biased with a small current $I_{B0}$ (this current is set to be 10 times smaller than $I_{B2}$). The size of these transistors is adjusted so that its $V_{GS}$ voltage is large enough to guaranty that transistors $M_1$ and $M_B$ are in the saturation region. The contribution of these bias transistors to the output noise power can be determined in a similar way as the noise contribution from the $R_s$ resistor (see Section 3.3.7). Assuming that $gm_{B2}/C_{gs} << \tau$ applies, the input referred noise due to transistor $M_{B2}$ is given by:

$$V^2_{N\,MB2in} \approx \frac{2 \times K_B \cdot T \cdot gm_{B2}}{\tau} \tag{5.5}$$

Therefore, this transistor is designed to have a transconductance value as small as possible, so that its contribution to the output noise voltage can be negligible when compared to the other noise sources in the circuit (this is achieved by making $I_{B0} \approx 0.1 \times I_B$ and $V_{DsatMB} > V_{DsatM1}$).

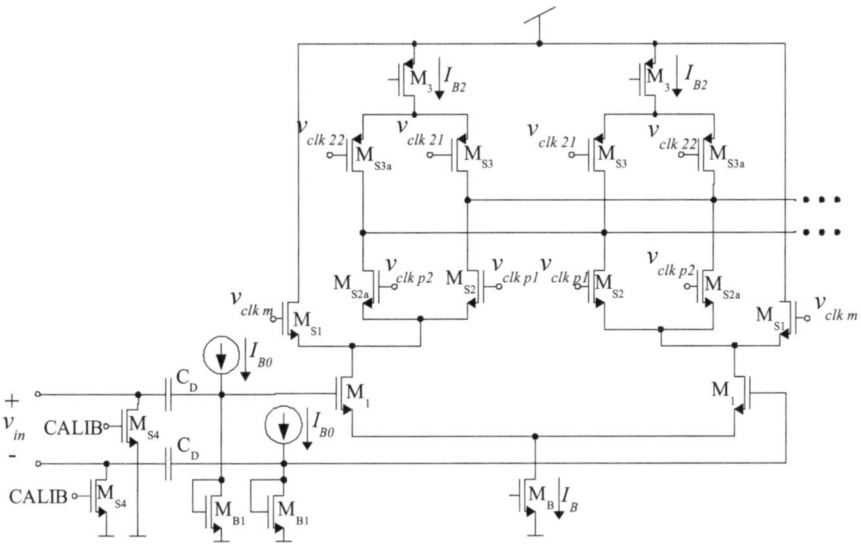


**Fig. 5.12** Differential Switch and active load circuit schematic

The decoupling capacitors $C_D$ introduce a zero at DC and an extra pole in the signal path. The input network of the circuit has two poles (due to $C_{gs1}$ and $C_D$) and a zero, forming a bandpass filter. The value of the capacitance $C_D$ should be large so that the low frequency pole of this network is mainly determined by the value of $C_D$ and that its frequency value is very low. The high frequency pole of the input network is determined both by the value of the $C_{gs1}$ capacitance and by the value of the source resistance, as was already shown in Chapter 3. The value of the capacitance of $C_D$ should be larger than the value of the $C_{gs1}$ capacitance, in order to have minimum attenuation of the input signal.

The expected supply currents of the sub-circuits of the switched-integrator (for the typical case) are shown next in Table 5.2.

**Table** 5.2 Expected supply current of the switched-integrator sub-circuits

| Circuit | Differential multiplier | Bias transistors | Load & CMF | Total |
|---|---|---|---|---|
| Current | 2000 μA | 200 μA | 400 μA | 2600 μA |

The interaction between the switched integrator and the amplifier circuit can be represented by the simplified equivalent single-ended circuit, shown on the left side of Fig. 5.13.

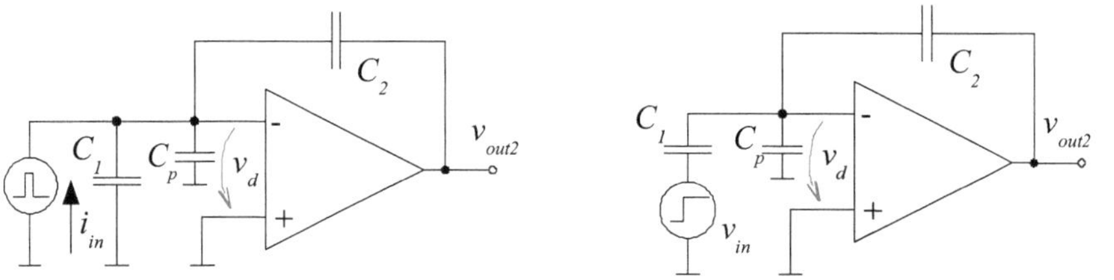


**Fig. 5.13** Simplified equivalent single-ended circuit for the amplifier with a current input (left side). Simplified equivalent single-ended circuit for the amplifier with a voltage step input (right side)

The switched-integrator circuit drives a current into the $C_1$ capacitor, during a time interval with duration equal to $\tau$ (this occurs once in every sample clock cycle). The op-amp has a finite *GBW* value and therefore it will take some time before its output voltage starts to change. If the value of $\tau$ is very small, when compared to the inverse of the op-amp *GBW* value, the switched integrator will produce a voltage step across $C_1$. In this case it is easier to use the circuit depicted in the right side of Fig. 5.13 to calculate the output voltage of the amplifier. Although the op-amp is connected in a shunt-shunt feedback configuration, it is easier to analyze the circuit assuming a voltage feedback. The equivalent block diagram of this circuit is shown in Fig. 5.14.

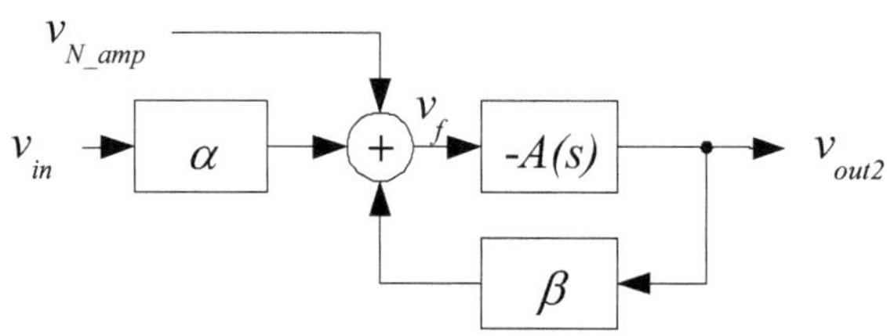


**Fig. 5.14** Block diagram of the amplifier feedback configuration

In the following analysis it will be considered that the op-amp has a dominant pole and its voltage transfer function ($A(s)$) can be approximated by a single pole transfer function:

$$A(s) \approx \frac{A_0}{s/p+1} \tag{5.6}$$

where $A_0$ is the DC gain of the op-amp and $GBW \approx A_0 \cdot p$. The feedback factor $\beta$ is given by:

$$\beta = \left[\frac{v_f}{v_{out2}}\right]_{v_{in}=0} = \frac{C_2}{C_1 + C_2 + C_p} \tag{5.7}$$

and the feed-forward factor $\alpha$ is given by:

$$\alpha = \left[\frac{v_f}{v_{in}}\right]_{v_{out2}=0} = \frac{C_1}{C_1 + C_2 + C_p} \tag{5.8}$$

finally, the closed loop transfer function of the amplifier circuit is given by:

$$\left[\frac{v_{out2}}{v_{in}}\right]_{v_{N\,amp}=0} = \frac{-\alpha \cdot A(s)}{1 + \beta \cdot A(s)} = -\frac{\alpha}{\beta} \cdot \frac{A_0}{1/\beta + A_0} \cdot \frac{1}{s/\left(p \cdot \left(A_0 \cdot \beta + 1\right)\right) + 1} \tag{5.9}$$

Note that if the op-amp is ideal (infinite DC gain and infinite *GBW*) the previous expression simplifies to $-C_2/C_1$, as expected. The output voltage as function of time, for a step input with an amplitude $V_{in}$, can be obtained by calculating the inverse *Laplace* transform of the previous expression multiplied by $V_{in}/s$., resulting in:

$$v_{out2}(t) = -V_{in} \cdot \frac{\alpha}{\beta} \cdot \frac{A_0}{1/\beta + A_0} \cdot \left(1 - e^{-t \cdot p \cdot (A_0 \cdot \beta + 1)}\right) \tag{5.10}$$

The previous expression shows that the output voltage of the circuit differs from the ideal output voltage value $-V_{in} \cdot C_2/C_1$. The actual output voltage value, at the end of the sampling period ($T_{SAMPLE}$), is smaller than the ideal value because of the finite DC gain value (second term in expression 5.10) and because of the finite $T_{SAMPLE}/GBW$ value (third term in expression 5.10). The relative error (settling error) between the desired output voltage and the real output voltage can be reduced by increasing the DC gain value and the *GBW* value of the op-amp. The minimum required values for the DC gain and GBW of the op-amp in order to obtain a relative settling error smaller than *error* are given by:[1]

[1] The following approximation was used: $A_0 \cdot p \cdot \beta + 1 \approx GBW \cdot \beta$.

$$A_0 > \frac{1-error}{error} \cdot \frac{1}{\beta}$$
$$GBW > \frac{\ln(error)}{\beta \cdot T_{SAMPLE}} \tag{5.11}$$

The op-amp creates noise that will appear in the output voltage of the circuit. The contribution of the op-amp noise to the output voltage can be calculated from the input referred noise voltage of the op-amp ($V_{N\,amp}$) considering that the amplifier has first order transfer function. The PSD of the output noise voltage can be calculated using the circuit diagram in Fig. 5.14. resulting in:

$$S_{N\,out}(f) = S_{N\,amp} \cdot \left| \frac{A(s)}{1+\beta \cdot A(s)} \right|^2 = S_{N\,amp} \cdot \frac{A_0^2}{4 \cdot \pi^2 \cdot f^2 / p^2 + (A_0 \cdot \beta + 1)^2} \tag{5.12}$$

where $S_{N\,out}$ and $S_{N\,amp}$ are the PSD of the output noise voltage and of the input referred noise voltage. The total output noise voltage due to the op-amp is given by (assuming that $S_{N\,amp}$ is constant):

$$V_{N\,out}^2 = \int_0^\infty S_{N\,amp} \cdot \frac{A_0^2}{4 \cdot \pi^2 \cdot f^2 / p^2 + (A_0 \cdot \beta + 1)^2} df = \frac{S_{N\,amp} \cdot A_0^2 \cdot p}{4} \tag{5.13}$$

It is important to note that the previous expression does not depend on the feedback factor ($\beta$) and therefore the closed loop gain value of the amplifier circuit does not affect the output noise level of the circuit. This occurs because when the closed loop gain decreases, the closed loop bandwidth of the amplifier increases, resulting in a constant value for the output noise voltage of the amplifier.

The voltage gain value of this circuit is set to 10, meaning that the value of $C_2$ capacitance is set to 1 pF, resulting in $\alpha \approx 1$ and $\alpha \approx 0.1$ (the value of $C_p$ is much smaller than the value of $C_1$). The *GBW* and DC gain of the amplifier were selected to obtain a settling error inferior to 0.5%, which results in $A_0$ >3990 (~72 dB) and *GBW* > 23.85 MHz. The amplifier will be designed to drive a load capacitance with value equal to 20 pF in order to facilitate the experimental evaluation of the circuit.

The topology selected for the amplifier is a two stage amplifier with a half folded-cascode first stage [1], using cascode compensation [2]. This topology allows for a large voltage swing while using a low power supply voltage value and for a high DC gain. The common mode feedback circuit uses two resistors to sense the common mode output voltage; it then compares this voltage with a reference voltage in a differential pair and adjusts the bias current of the top current source PMOS transistors. The schematics of both the amplifier circuit and the common-mode feedback circuit are shown in Fig. 5.15.

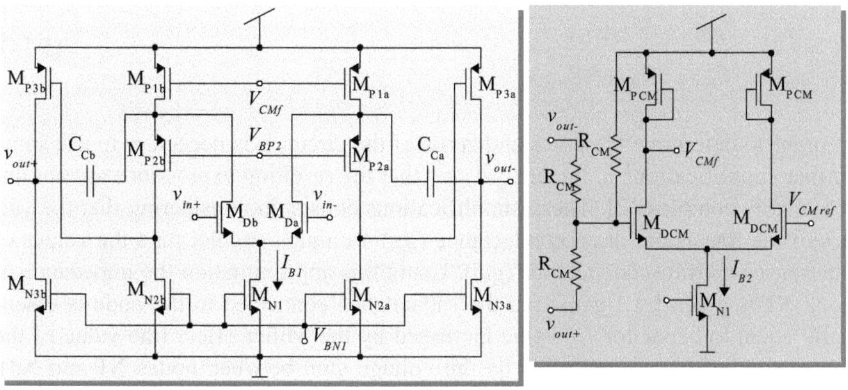


**Fig. 5.15** Schematics of the amplifier and amplifier common-mode feedback circuits

The small signal transfer function of the op-amp can be calculated using a simplified single-ended version of the previous schematic, shown in Fig. 5.16.

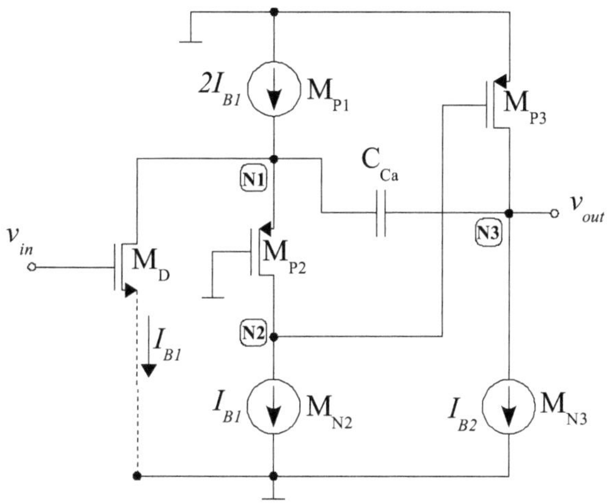


**Fig. 5.16** Simplified single-ended version of the op-amp schematic

Using the previous circuit it is possible to calculate the op-amp voltage transfer function, assuming that the signals have a small amplitude. The DC gain of the amplifier is the product of the gains of the first and second stage. The gain of the first stage is the product of the transconductance of the input transistor ($M_D$) and the impedance of the node **N2.** The gain of the second stage is the product of the transconductance of the transistor $M_{P3}$ and the impedance of node **N3**. Since the drain conductance of transistor $M_{N2}$ is much larger than the conductance of the cascode circuit formed by transistors $M_{P2}$, $M_{P1}$ and $M_D$, the expression of the DC gain can be simplified to:

$$A_0 = \frac{v_{out}}{v_{in}} \approx \frac{gm_D}{g_{ds\,N2}} \cdot \frac{gm_{P3}}{g_{ds\,P3} + g_{ds\,N3}} \tag{5.14}$$

In order to determine the poles and zeros of the circuit it is necessary to use some further simplifications in the analysis, so that the resulting expressions are not unreasonably complex [2]. These simplifications consist in considering that the values of the transistors drain conductance ($g_{ds}$) are much smaller than the values of the transistors transconductance ($gm$). Using this approximation the impedance of node **N1** is given by $1/gm_{P2}$ and the capacitance connected to the node is essentially equal to capacitor $C_C$ value increased by the Miller effect (the value of the capacitance of $C_C$ is multiplied by the voltage gain between nodes **N1** and **N3**). Therefore the value of the pole associated to node **N1** is given by:

$$p_1 \approx -\frac{gm_{P2}}{C_C \cdot G_{V13}} = -\frac{gm_{P2}}{C_C \cdot \frac{gm_{P2}}{g_{ds\,N2}} \cdot \frac{gm_{P3}}{g_3}} = -\frac{g_{ds\,N2}}{C_C \cdot \frac{gm_{P3}}{g_3}} \tag{5.15}$$

where $g_3$ is the conductance associated with node **N3** (equal to $g_{dsP3}+g_{dsN3}$). The poles associated with nodes **N2** and **N3** are complex conjugated, due to the local feedback loop created by $C_c$, with a natural frequency and quality factor given by [2]:

$$\omega_{p\,1,2} \approx \sqrt{\frac{gm_{P2} \cdot gm_{P3}}{C_L \cdot C_{p2}}} \tag{5.16}$$

$$Q_{p\,1,2} \approx \sqrt{\frac{gm_{P3} \cdot C_L}{gm_{P2} \cdot C_{p2}}} \cdot \frac{C_C}{C_L + C_C} \tag{5.17}$$

where $C_{p2}$ is the parasitic capacitance of node **N2**, $C_C$ is the compensation capacitor and $C_L$ is the load capacitance connected to node **N3**. Examining expressions 5.15 and 5.16 leads to the conclusion that the dominant pole of the circuit is the pole $p_1$. Assuming that the amplifier is stable (making 5.16 at least two times larger than 5.15), the *GBW* of the circuit is given by:

$$GBW \approx p_1 \cdot A_0 = \frac{g_{ds\,N2}}{C_C \cdot \frac{gm_{P3}}{g_3}} \cdot \frac{gm_D}{g_{ds\,N2}} \cdot \frac{gm_{P3}}{g_{ds\,P3} + g_{ds\,N3}} = \frac{gm_D}{C_C} \tag{5.18}$$

In order to have a stable amplifier it is also important that the quality factor of the complex poles (expression 5.17) is smaller than 1. The input signal can travel from node **N1** to the output node **N3** through two different paths, one path through transistors $M_{P2}$ and $M_{P3}$ and the other path through the capacitor $C_c$. These two signal paths cause two zeros in the amplifier transfer function. One zero is positive and the other is negative, these zeros have approximately the same frequency value (in modulus), given by

$$\left|z_{1,2}\right| \approx \sqrt{\frac{gm_{P2} \cdot gm_{P3}}{C_C \cdot C_{p2}}} \tag{5.19}$$

this frequency value can be (by design) larger than any pole frequency value in the circuit, therefore these zeros do not influence significantly the stability of the amplifier. Since the specification for the output noise voltage of the amplifier is not stringent (the output noise voltage should be smaller than 10 mVrms), the noise performance of the amplifier is not an important design parameter.

The amplifier is designed to be stable and to have a DC gain value larger than 72 dB and a GBW value larger than 24 MHz for a 20 pF load. The design criteria for this circuit is as follows

- The value of $C_C$ is selected to be at least 10 times smaller than $C_L$ to reduce the value of quality factor of the complex poles.
- $M_{P2}$ this transistor is designed to have a large *gm* value in order to increase the frequency of the complex poles and to reduce its quality factor values. This can be achieved by using a small $V_{Dsat}$ value. The channel length of this transistor is set to minimum length in order to reduce the parasitic capacitance in node **N2**.
- $M_{P3}$ this transistor is designed to have a *gm* smaller than transistor $M_{P2}$ in order to reduce the value of the quality factor of the complex poles. The $V_{Dsat}$ value of this transistor is a compromise between the output voltage swing and the value of *gm*. The channel length value is obtained from a compromise between the parasitic capacitance in node **N2** and the value of the DC gain.
- The value of the current $I_{B1}$ is higher than the value of the current $I_{B2}$ in order to obtain a smaller *gm* value in transistor $M_{P3}$ when compared to transistor $M_{P2}$.
- $M_{N2}$ this transistor is designed to have a large $V_{Dsat}$ value in order to reduce the parasitic capacitance in node **N2**, the channel length value is large in order to increase the DC gain.
- $M_{N3}$ this transistor is designed to have a large channel length value in order to increase the DC gain.
- $M_D$ This transistor is designed to have a $V_{Dsat}$ value large enough to obtain the desired GBW value.

The output noise voltage of the amplifier is determined by an electrical simulation of the circuit to be around 4 mVrms. The expected supply currents of the sub-circuits of the amplifier (for the typical case) are shown next in Table 5.3.

**Table 5.3** Expected supply current of the amplifier sub-circuits

| Circuit | amplifier | Bias transistors | CMF | Total |
|---|---|---|---|---|
| Current | 800 μA | 20 μA | 80 μA | 900 μA |

## *5.1.3 Prototype Integrated Circuit*

The main objective of the prototype integrated circuit is to test the sub-blocks and circuits that constitute the radar architecture. Although the prototype can be considered a complete UWB radar transceiver integrated circuit, it has more pins than it would be necessary to work just as a transceiver circuit. These extra pins are necessary to provide access to internal nodes, to adjust the bias currents of the different sub-circuits and to adjust the reference voltages needed for (almost) every sub-circuit. In a final transceiver integrated circuit design, all these reference voltages and bias currents values would be fixed and generated internally. There is a TEST signal that allows configuring the transceiver into a test mode. In this mode it is possible to observe the output clock of the DLL and to provide an independent sample clock to the switched-integrator circuit. This allows testing each of these blocks independently. This test scheme is shown in Fig. 5.17.

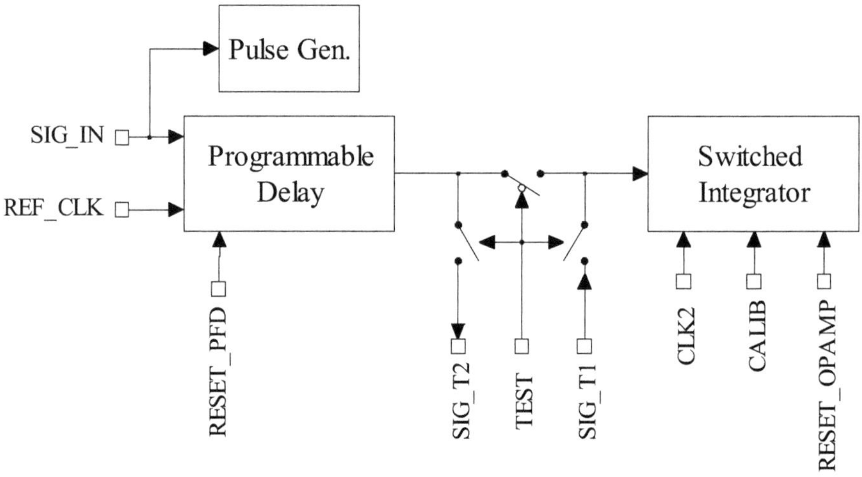


**Fig. 5.17** Test scheme of the transceiver prototype

The digital input-output digital signals of the transceiver prototype are listed in Table 5.4

**Table 5.4** List of the digital input output signals of the transceiver prototype

| Signal | Description |
|---|---|
| RESET_OPAMP | Control signal of the reset switches of the op-amp feedback capacitors |
| SIG_IN | Input clock signal of the programmable delay circuit (PRF clock) |
| SIG_T1 | When TEST=1 this signal is the sample clock of switched-integrator |
| TEST | Test mode control signal |
| SIG_T2 | When TEST=1 this signal is the output clock of the programmable delay |
| REF_CLK | Input reference clock of the programmable delay (generated by the ΣΔ) |
| RESET_PFD | Reset signal for the PFD circuit |
| CALIB | Control signal of the switched-integrator calibration mode |
| CLK2 | Control signal of the active load in the switched-integrator |

The floor plan diagram of the integrated circuit is shown in Fig. 5.18. The largest block is the DLL circuit (programmable delay), this is because of the large capacitance needed for the loop filter. The receiver channel is in the middle of the circuit with the input pins located on the bottom and the output pins located on the top. This configuration provides the maximum possible distance between the input and output voltages and thus limits the possibility of unwanted oscillations (due to a parasitic feedback path) to occur.

In the floor plan diagram, all the analog input and output pads are shaded in gray, all the digital pads have a pattern formed by "Ds", all the positive power supply pads have rectangular pattern and all the ground pads have a cross pattern. There are two pairs of power supply pads for the digital circuits: a 1.5 V power supply for the internal digital circuits and a 3.3 V power supply for the interface pads (these pads realize a voltage level conversion). There are three pairs of analog 1.5 V power supply pads for the analog circuits; one pair for the short pulse generator circuit, one pair for the receiver channel circuit and one pair for the programmable delay circuit.

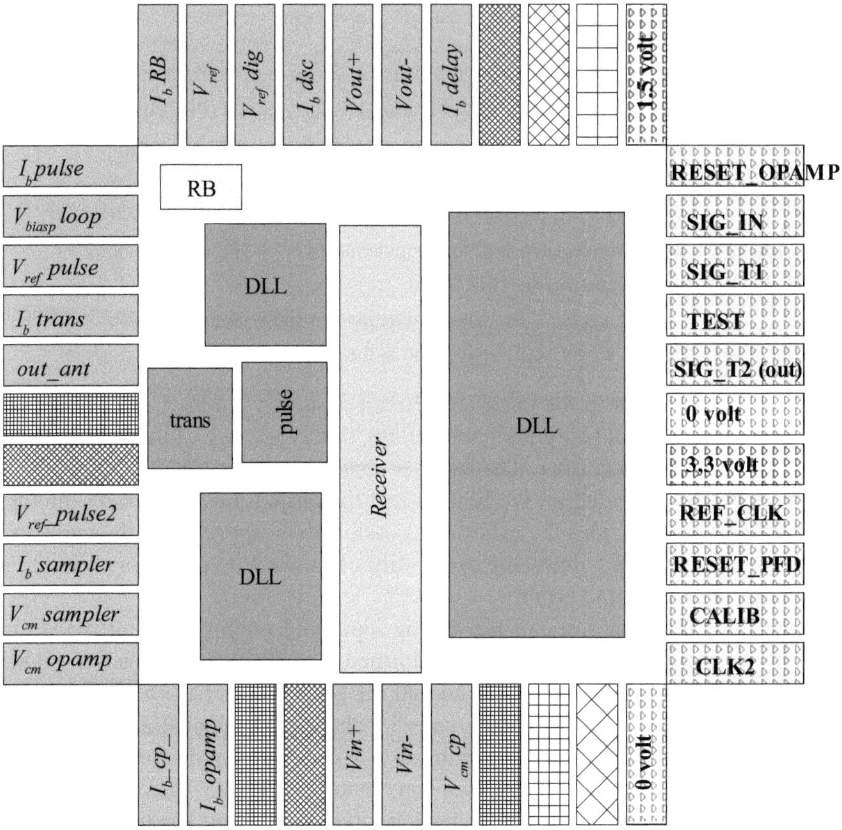


**Fig. 5.18** Floor plan diagram of the prototype integrated circuit. The digital pads are marked with the pattern "D"

The digital pads must be grouped together, separated from the analog pads, because each type of pad must have its own separate ESD bus, with separated power supply pins. The two ESD buses are separated by breaker pads, located between the 0 V analog power pad and the 1.5 V digital power pad (on the top) and located between the 1.5 V analog power pad and the 0 V digital power pad (located at bottom). The digital inputs of the circuit are located on the right side of the circuit and for that reason the DLL circuit (which is the circuit that uses the largest number of digital inputs) is also located on the right side. When necessary, each digital pad is connected to either a CMOS to differential converter circuit (digital input pad) or to a differential to CMOS circuit (digital output pad).

The capacitors of the DLL loop filter occupy a large area and therefore are distributed across the integrated circuit. All the high frequency signals are located as close as possible to the corresponding pads. The short pulse generator circuit, that produces the output pulse, is located in the left side and is located very close to its power supplies and output pads. The receiver channel inputs are also very close to the input pads.

The layout of the integrated circuit is shown in Fig. 5.19. The selected CMOS technology has a minimum channel length of 0.18 μm, 6 layers of metal for interconnecting the various components and Metal to Metal capacitors (MiM ). The location (and area) of the blocks is the same as in the floor-plan diagram. Most of the area of the layout is occupied by the capacitors of the circuit. As already mentioned, the capacitors of the DLL circuit are distributed. The digital circuits of the DLL (phase comparator and VCDL) are located on the top and the analog circuits (charge pump and differential to single ended converter) are located on the bottom of the DLL bock. The receiver channel has the input switches on the bottom, followed by the AC coupling capacitors, the differential switch, the $C_1$ capacitors and finally, on the top, the amplifier. The power supply lines have a large width to reduce its resistance and whenever possible, decoupling capacitors were added in parallel with the power supplies lines to improve the stability of the power supply voltage.

The complete layout of the pads and corresponding circuit schematic was unavailable, the layout of the test circuit used an "abstract" view which only indicates where to connect the desired input or output signals. This abstract view is replaced by the actual layout of the pad before the fabrication of the circuit. As a consequence there was a layout error in the test circuit: the input signal of the switched-integrator circuit was connected to the corresponding pads using metal 6 lines (the top metal layer). However the pads only use the first 4 layers of metal. Therefore the input signal was not connected and it was impossible to measure the performance of the switched-integrator circuit. This error was later corrected in some dies using a Focused Ion Beam machine (FIB), this process allows to remove the passivation oxide and metal layers and to deposited new metal in the circuit, thus reestablishing the connection between the metal 6 in the input lines of the switched-integrator circuit and the metal 4 of the input pads.

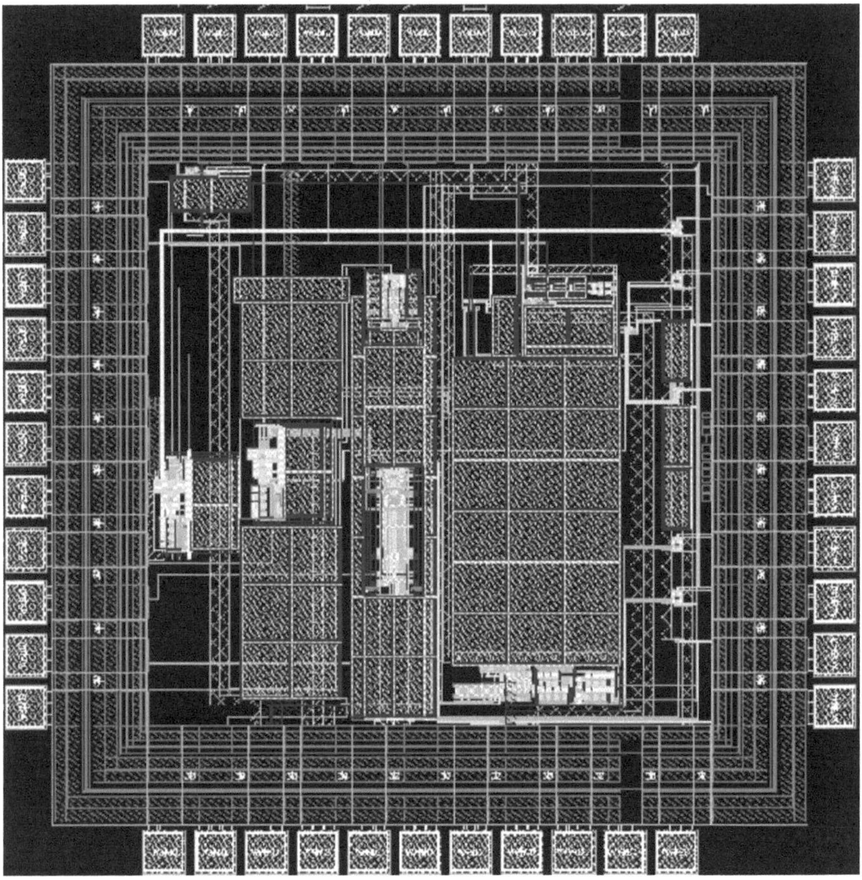

**Fig. 5.19** Layout of the prototype integrated circuit

## 5.2 Experimental Results

In order to evaluate the performance of the different sub-circuits, it is necessary to build a PCB to supply all the voltages and currents necessary for the circuit. This PCB has four layers, where two are connected to ground, in order to reduce the coupling from the digital signals into the power supplies and the references. Each reference voltage and current can be adjusted independently using resistive trimmers. The test board has four separate power supplies: two 1.5 V power supplies for the IC analog and digital power supplies, one 2.7 V power supply for the board reference voltage generators and one 3.3 V power supply for the IC digital power supply. All these power supplies are decoupled using several capacitors in parallel and an inductor in series. Each power supply pin of the IC is also decoupled by a capacitor placed under the IC. In order to reduce the size of the board and place the components as close as possible to the IC, most of the components in the test board are SMD.[2] A photograph of the test PCB is shown in Fig. 5.20.

Fig. 5.20 Photograph of the test PCB

In order to reduce the inductance of the bond-wires (this inductance is approximately 1 nH/mm [3]), the silicon die is direct bonded to the PCB. This solution results in very small bond wires (the length of the wires is smaller than 2 mm). If the circuit is encapsulated in a package, it must use a 44 pin package. These packages have 11 pins on each side which results in a cavity much larger than the silicon die of the prototype IC. The result is that the bond wires connecting the circuit pads to the package have a length around 4 mm. The pin structure in the package also adds parasitic capacitance and inductance [4]. Photographs of the silicon die glued to the test PCB (direct bonding) and of the silicon die in a JLCC package, are shown in Fig. 5.21, to illustrate this point. The effects of the parasitic components of the bond-wire connection in the signal are discussed in Appendix D.

[2] SMD: Surface Mounted Device.

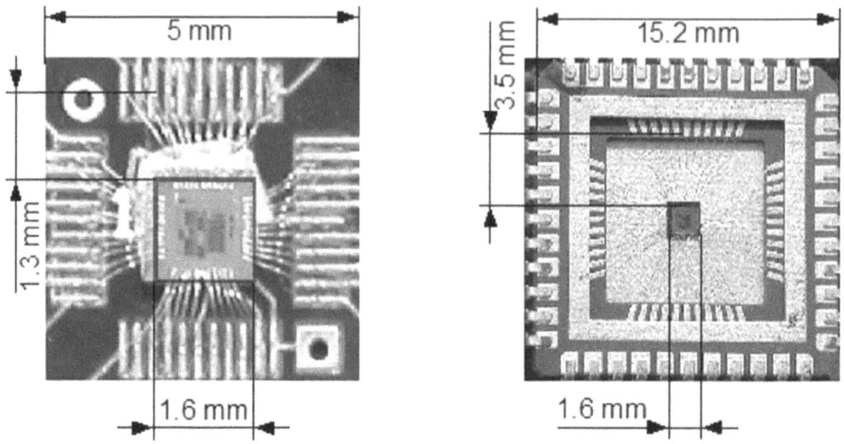


**Fig. 5.21** Photographs of the silicon die on the PCB and in a JLCC package

The diagram of the experimental test set-up is shown in Fig. 5.22. The power supply voltages are provided by a programmable power supply (Tektronix PS2521G) with current limiting. The digital signals are created by a digital pattern generator (Tektronix AWG 510). The output signals are observed and digitized using a digital oscilloscope (Tektronix TDS3052). A digital multimeter (Agilent 34401A) is used to measure the values of the DC currents and voltages in the circuit. The input sine wave signals are generated by a network analyzer (HP 8753ES).

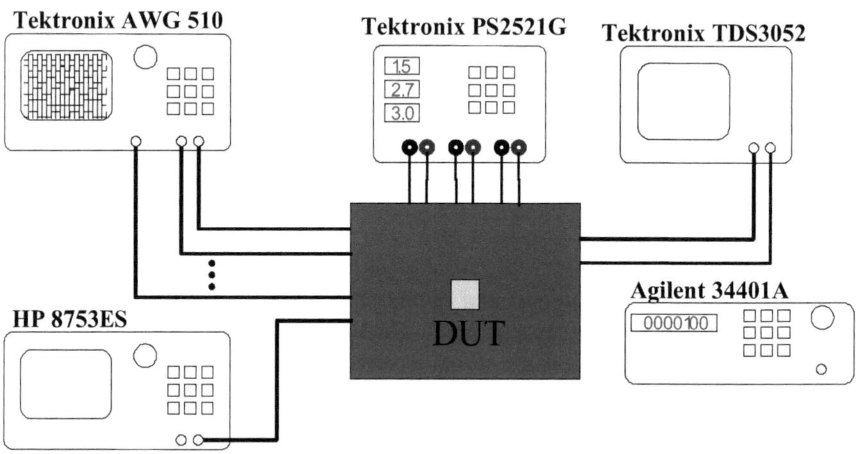


**Fig. 5.22** Diagram of the experimental test set-up

The nominal and measured values of the reference currents of the prototype are shown in Tables 5.5 and 5.6.

**Table 5.5** Bias currents of the prototype circuit

| Pin Name | Nominal value | Measured value | Description |
|---|---|---|---|
| $I_B$ RB | 100 μA | 100 μA | Bias current of the master replica bias circuit |
| $I_B$ dsc | 10 μA | 7 μA | Bias current of the differential to single ended-circuit |
| $I_B$ del | 100 μA | 100 μA | Bias current of the replica bias circuit of the VCDL |
| $I_B$ op-amp | 5 μA | 5 μA | Bias current of the op-amp |
| $I_B$ CP | 80 μA | 80 μA | Bias current of the charge pump circuit |
| $I_B$ sampler | 100 μA | 100 μA | Bias current of the switched-integrator |
| $I_B$ trans | 100 μA | 100 μA | Bias current of the replica bias circuit of the output short-pulse generator |
| $I_B$ pulse | 100 μA | 100 μA | Bias current of the replica bias circuit of the short-pulse generator of the switched-integrator |

**Table 5.6** Reference voltages of the prototype circuit

| Pin Name | Nominal value | Measured value | Description |
|---|---|---|---|
| $V_{ref}$ dig | 1.3 V | 1.3 V | Reference voltage used in the CMOS to differential converter circuit $V_{ref}dig = \frac{V_{DD} - V_{ref}}{2}$ |
| $V_{ref}$ pulse | (0.2 – 0.8) V | (0.2 – 0.8) V | $V_{BiasP}$ reference voltage of the replica bias circuit of the output short-pulse generator |
| $V_{CM}$ op-amp | 1 V | 0.9 V | Reference voltage for the op-amp common mode feedback. |
| $V_{ref}$ | 0.9 V | 0.9 V | Reference voltage of the differential logic circuits. |
| $V_{ref}$pulse2 | (0.2 – 0.8) V | (0.2 – 0.8) V | $V_{BiasP}$ reference voltage of the replica bias circuit of the short-pulse generator of the switched-integrator |
| $V_{CM}$ sampler | 1 V | 1.1 V | Reference voltage for the active load circuit in the switched-integrator. |
| $V_{CM}$ CP | 0.8 V | 1.05 V | Reference voltage for the charge pump common mode feedback circuit. |

In the previous Tables, the column with the nominal values corresponds to the voltage or current values determined in the project of the circuits. The column with the measured values corresponds to the voltage or current values that produced the best results in the experimental evaluation of the prototype circuits. The measured analog power supply current dissipated by the circuit is equal to 5.4 mA, for a power supply voltage value equal to 1.5 V.

### *5.2.1 Short Pulse Generator*

In order to measure the output voltage of the short-pulse generator circuit, a 2.5 MHz CMOS clock signal with 50% duty cycle is applied to the SIG_IN input of the circuit. The output voltage of the circuit is measured using the digital oscilloscope (TDS 3052), this oscilloscope is also used to measure the pulse width and amplitude. The input impedance of the oscilloscope is composed by a 13 pF capacitor in parallel with a load resistance of 50 Ω or 1M Ω, the output signal of the short pulse generator was measured for different values of the bias voltage ($V_{ref}$ *pulse*) and for the two load resistance values. As an example, the output voltage waveform of the short pulse generator for a bias voltage ($V_{ref}$ *pulse*) equal to 300 mV and a load resistance equal to 1 MΩ, is shown in Fig. 5.23.

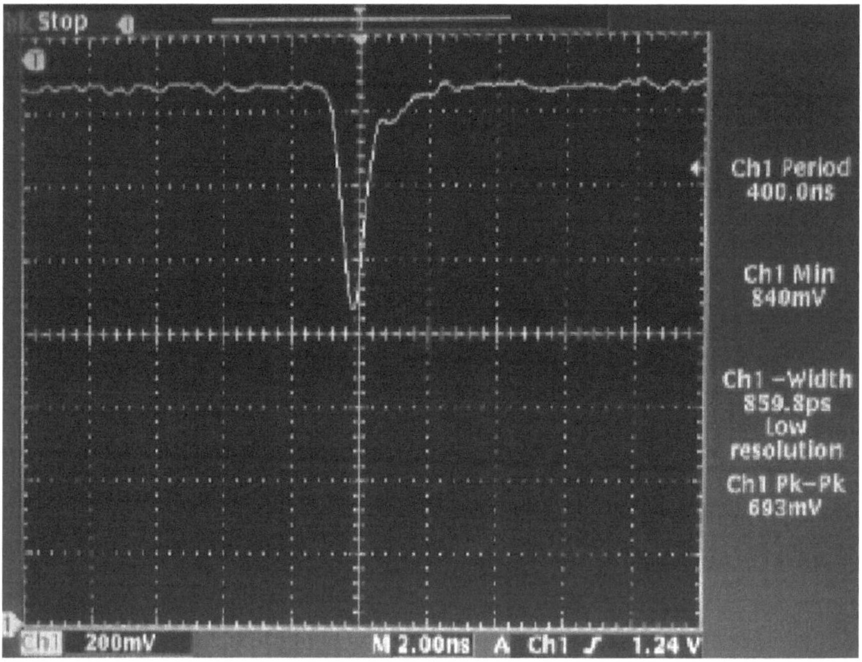


**Fig.** 5.23 Output voltage of the short pulse generator circuit for $V_{ref\,pulse}$=0.3 V and a load resistance value of 1 Mohm

The measured output voltage has a similar shape as the simulated output voltage of the short pulse generator circuit (see Fig. 5.4). It should be noted that the bandwidth of the oscilloscope is only 500 MHz; therefore the observed pulse shape is not exact.

The pulse width and amplitude was measured using the oscilloscope, for different bias voltages values, the resulting values, as a function of the bias voltage and load resistance values are shown in Figs. 5.24. and 5.25.

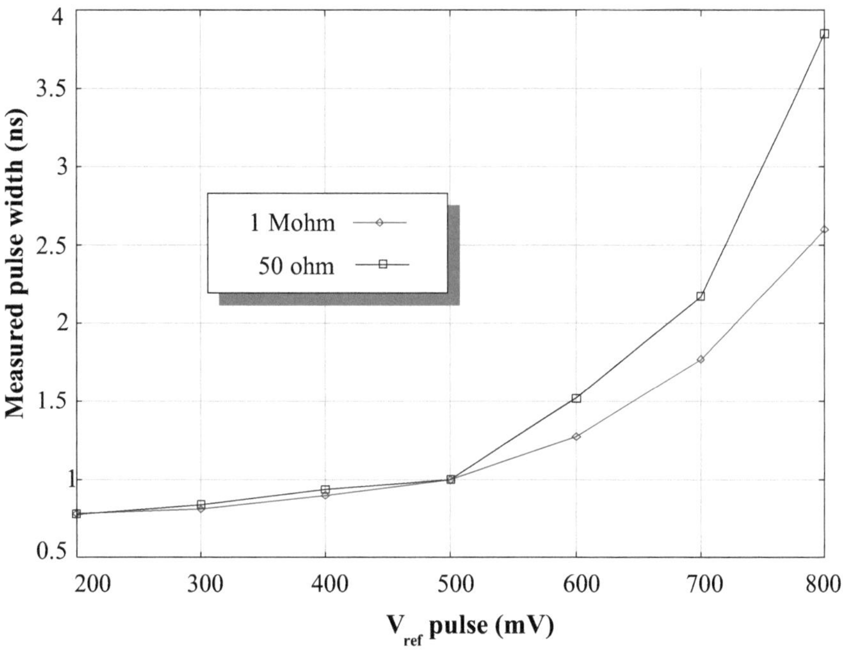


**Fig. 5.24** Measured pulse width as a function of the bias voltage for two different load values

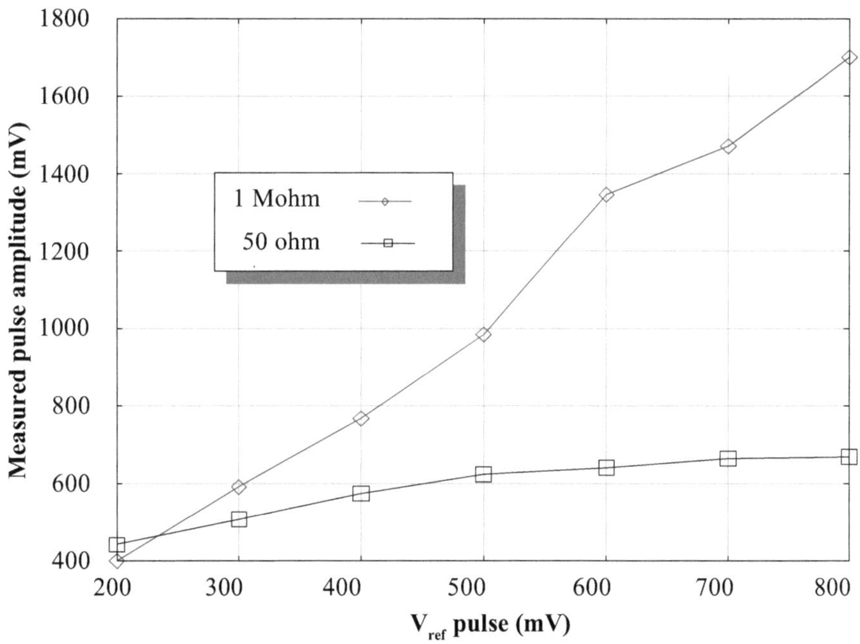


**Fig. 5.25** Measured pulse amplitude as a function of the bias voltage for two different load values

The previous graphs show that the short-pulse generator circuit is capable of creating UWB short pulses with time duration between 0.7 ns and 3 ns. The pulse amplitude changes both with the control voltage and the load of the circuit. The amplitude of the pulse is smaller in the case of the 50 ohm load. These results are in agreement with the simulation results of the circuit.

Since the oscilloscope creates a vector with the sampled input voltages, the PSD of the output pulse can be estimated applying the FFT to the vector containing the pulse voltage sampled with a sampling frequency of 5 GHz. The resulting PSD of the output pulse when the $V_{BiasP}$ is equal to 0.6 V and the load resistance is 50 ohms, is shown in Fig. 5.26.

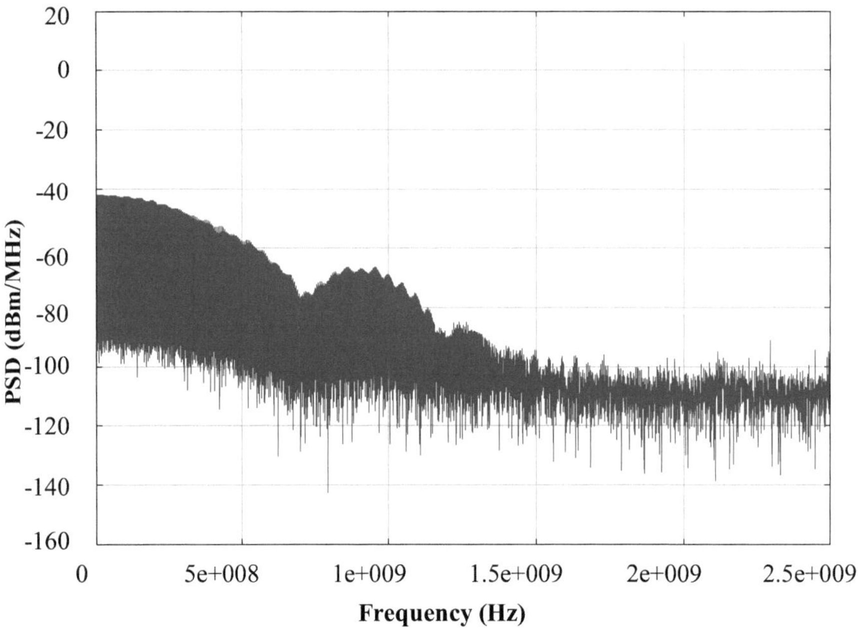


**Fig. 5.26** PSD of the output pulse for $V_{BiasP}$=0.6 V and a load resistance value equal to 50 ohms

## *5.2.2 Switched-Integrator*

In order to test the switched integrator circuit, the TEST signal is set to one, making the sampling clock of the switched integrator circuit the SIG_T1 clock signal. This signal together with the other clock signals of the switched integrator circuit is generated by the digital pattern generator (AWG 510).

The offset voltage of the switched integrator circuit was measured by making the circuit sample an input voltage equal to 0 V and then measuring the output voltage. In order to reduce the error in this measurement, the switched integrator adds 256 samples together, thus increasing the value of the output voltage. The number of samples added together is controlled by the RESET clock signal, when this signal is ON the amplifier output voltage is set to 0 V. The offset voltage was measured using the three different clock schemes depicted in Fig. 5.27. The sampling instant is determined by the rising edge of the SIG_T1 clock signal.

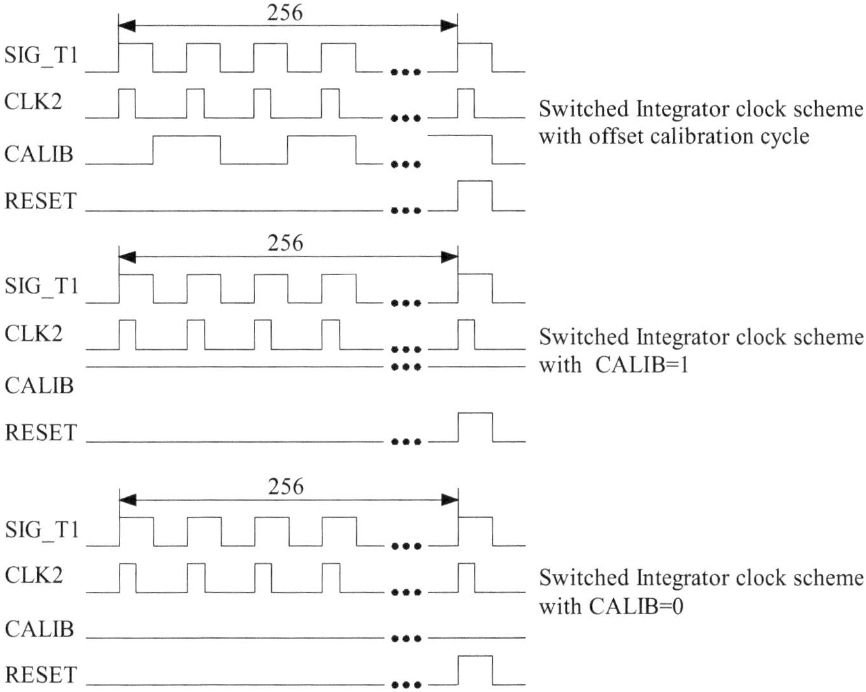


**Fig. 5.27** Switched-Integrator clock signals. From top to bottom: clock scheme with calibration cycle, clock scheme with CALIB=1 and clock scheme with CALIB=0

The measured output voltages of the switched integrator circuit, as a function of time, are shown in Fig. 5.28. The voltages are reseted to zero and during each sampling clock cycle a negative offset voltage is added, resulting in a saw-tooth waveform.

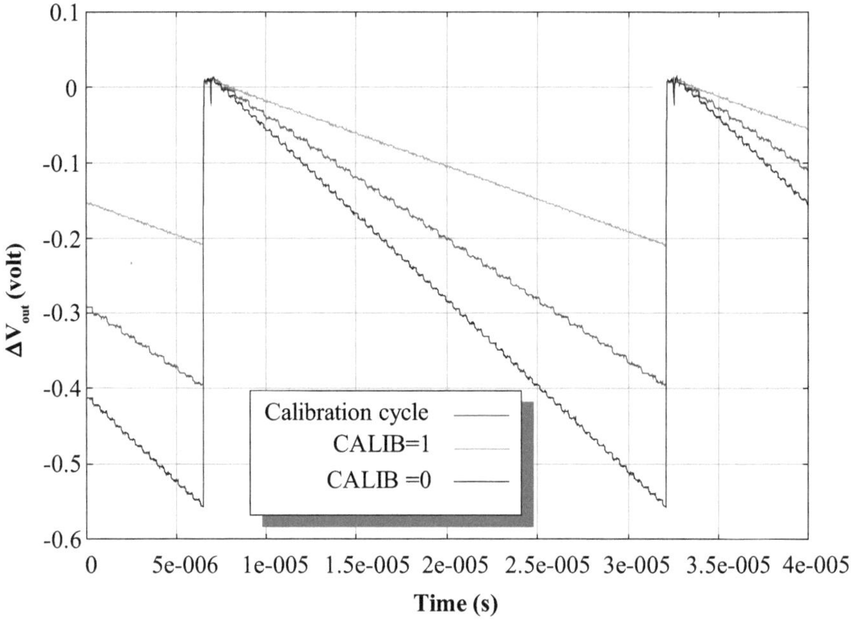


**Fig. 5.28** Measured differential output voltage with $\Delta V_{in}$=0 V

The output offset voltage values of the switched integrator after 256 sampling clock cycles are 390 mV with the calibration cycle, 196 mV with CALIB=1 and 620 mV with CALIB=0. These values show a problem with the offset calibration scheme since the output offset voltage is large, even when using the offset calibration cycle (390 mV). In fact the output offset voltage is larger when using the offset calibration scheme than when CALIB=1. In this case, the switched-integrator input is connected to ground through the transistors $M_{S4}$ (see Fig. 5.12). When CALIB=0 the input of the switched integrator is connected to ground externally (in the PCB board) and the polarity of its output differential current is inverted from the previous case (using switches $M_{S2}$ and $M_{S3}$). This means that if the offset voltage origin was inside the switched integrator circuit, the measured output voltage values should be symmetrical when CALIB=1 and CALIB=0. It should be noted that these measured output offset voltage values are still within the predicted maximum output offset voltage value (see Section 3.3.9 of Chapter 3). In order to gain some more information about the offset voltage it is important to observe the behavior of the circuit in each sampling clock cycle. In Fig. 5.29. the measured switched integrator output voltage is magnified around the time when the RESET signal is activated.

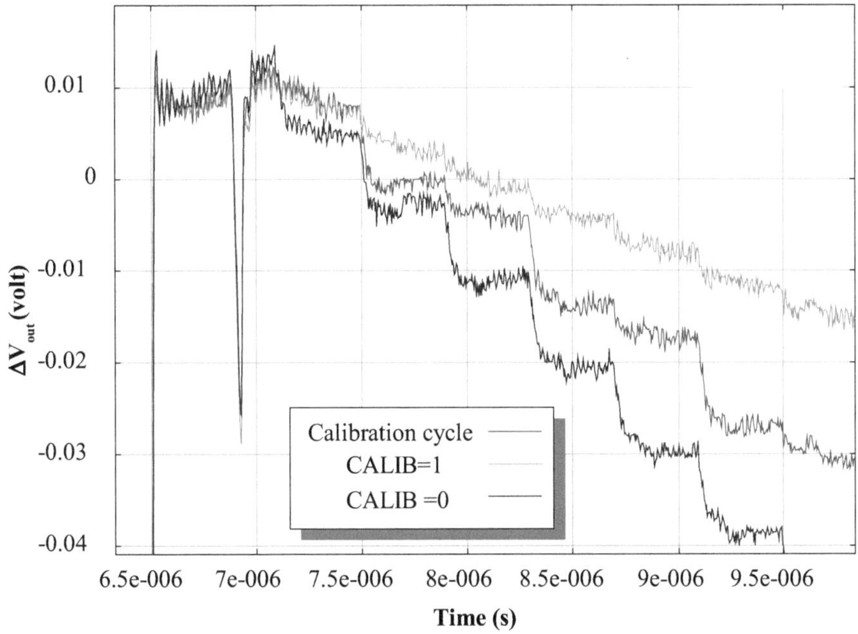


**Fig. 5.29** Measured differential output voltage with $\Delta V_{in}$=0 V (zoom)

The first thing to note in the previous graph is that the output voltage is reseted to approximately 9 mV instead of 0 V. This error is essentially due to the offset voltage in the oscilloscope, as explained in Appendix E. The graph shows that in each sampling cycle (with a period equal to 400 ns) the output voltage is decremented by a fixed step. This step is larger when the switched integrator is sampling the externally reseted input voltage (CALIB=0) than when it sampling the internally reseted input voltage (CALIB=1). In the case of the calibration cycle, the value of the offset step voltage alternates between the two previous step values, according to the CALIB clock signal value.

In order to measure the switched-integrator transfer function, a sine wave signal must be applied to the input of the switched integrator circuit. Since the input of this circuit is a differential voltage and the available signal generator (HP 8753ES) is only capable of generating a single ended sine wave signal it is necessary to connect one of the inputs of the switched integrator to ground. This is illustrated in Fig. 5.30.

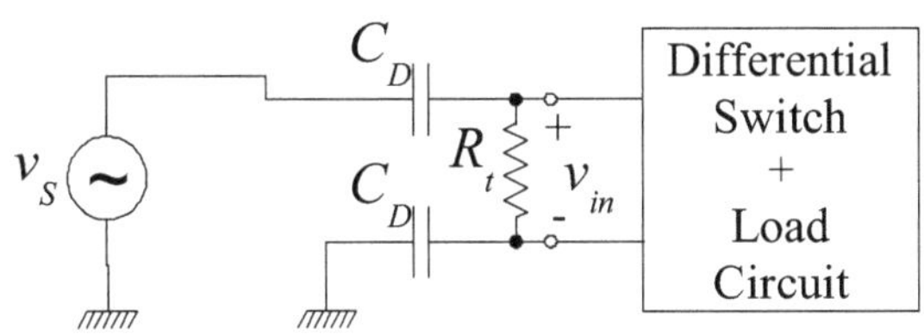


**Fig. 5.30** Experimental input signal set-up for the switched integrator circuit. The sine wave voltage source ($v_S$) is connected through a 50 ohm coaxial cable and terminated by a resistance $R_t$

It was preferred to use a single-ended signal to drive the circuit instead of using a transformer to create a differential signal, because it is very difficult to obtain a transformer with a wide-band frequency response and the objective is to measure the circuit transfer function as a function of the input signal frequency. In this case, using a single ended signal is not a problem, because the amplitude of the input signal is small and should not cause significant distortion in the circuit. The single ended input signal is transformed into a differential output voltage by the differential pair $M_1$ (see Fig. 5.12). The gain of the switched-integrator circuit is defined as the ratio of the amplitude of the differential output voltage to the amplitude of the input single-ended voltage signal. The output differential noise voltage of the circuit can be measurement disconnecting the voltage source and using the same experimental set-up. In this case the termination resistor $R_t$ works as the noise source of the input voltage source.

The gain of the circuit depends on the number of samples added together (see expression 5.2). However if a large value of samples are used the output voltage would be a distorted sine-wave, that would difficult the measurement of the output amplitude. Therefore the gain is measured using only one sample ($N$=1), in this case the offset is not integrated in the output voltage and it is possible to measure the output amplitude accurately. The clock scheme used in this measurement is shown in Fig. 5.31.

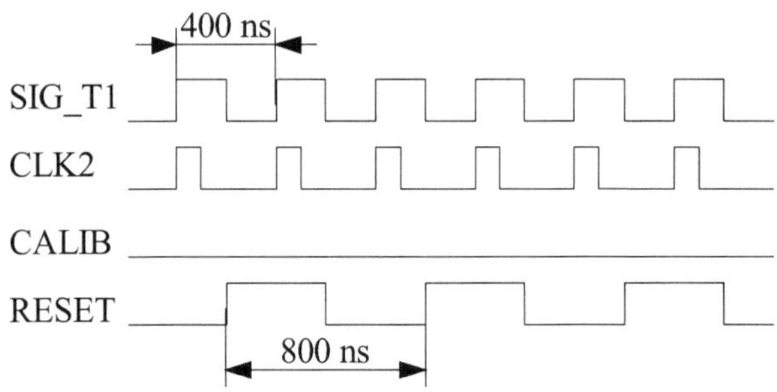


**Fig. 5.31** Switched-integrator clock waveforms for $N$=1

The switched-integrator when $N$=1 works essentially as a sub-sampling S&H circuit, it samples the input signal during the rising edges of the SIG_T1 clock signal that coincide with RESET=0 and the output voltage is set to zero volts when RESET=1. The sampling period is equal to 800 ns, corresponding to a sampling frequency equal to 1.25 MHz. The input signal frequency range is much larger than this value because the input signal is an UWB pulse that has a PSD with significant energy in frequencies up to 1 GHz. The frequency of the output signal of the circuit has always a frequency value between 0 Hz and $F_s/2$ (0.625 MHz), the exact value depends on the frequency of the input signal ($f_{in}$) and on the sampling frequency ($F_s$) of the switched-integrator circuit and is given by:

$$f_{out} = \left| f_{in} - n \cdot F_S \right| \tag{5.20}$$

where n is equal to 0, 1, 2 , ... and is determined in order for the output frequency value to be between 0 Hz and $F_s$ /2. In order to facilitate the measurement of the amplitude of the output voltage of the circuit, the frequency values of the input signal are selected in order for the frequency value of the output signal be always equal to 100 kHz. As an example, Fig. 5.32 shows the measured differential output voltage when the input signal frequency is equal to 100.1 MHz and the input amplitude is 400 mV.

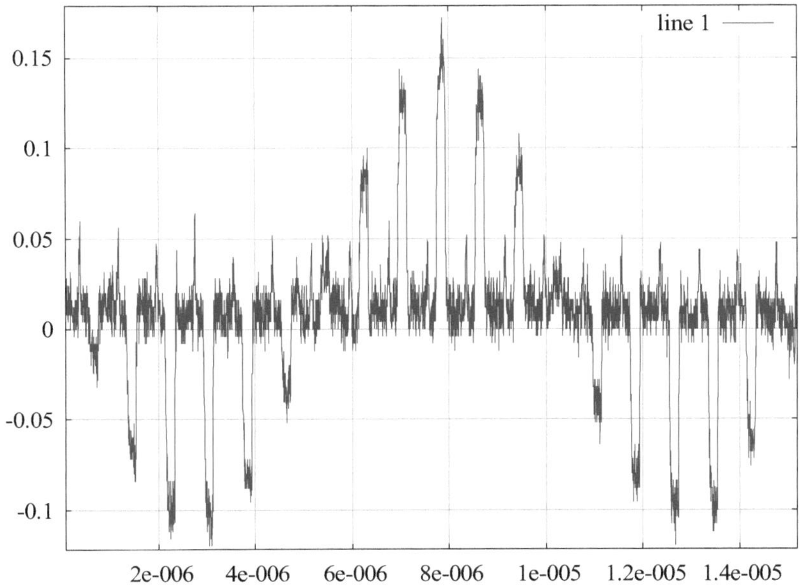


**Fig. 5.32** Measured Switched-integrator differential output voltage for an input signal with a frequency value equal 100.1 MHz an amplitude value equal to 400 mV

The gain of the circuit can be calculated by dividing the amplitude of the output signal by the amplitude of the input signal (this amplitude was measured with the oscilloscope in the input pin of the circuit). This calculation results in a voltage gain equal to 0.43 (–7.33 dB), which is in disagreement with the voltage gain value obtained trough the electrical simulation of the circuit (equal to 6.5 or 16.3 dB). The measured voltage gain of the circuit is 23.5 dB less than the simulated voltage gain; this is clearly a significant difference. A clue to where the problem is located is the shape of the output differential voltage; which is a sine wave shape. This should not be the case since the differential pair formed by the transistors $M_1$ has a $V_{Dsat}$ voltage equal to 120 mV and the input sine wave has an amplitude equal to 400 mV, therefore this differential pair should be clearly saturated for this input signal and produce a compressed sine-wave shape in the output voltage. This fact is confirmed by the simulated differential output voltage shown in Fig. 5.33. In this case the input signal is a sine-wave with a frequency value equal to 100.05 MHz and an amplitude equal to 300 mV.

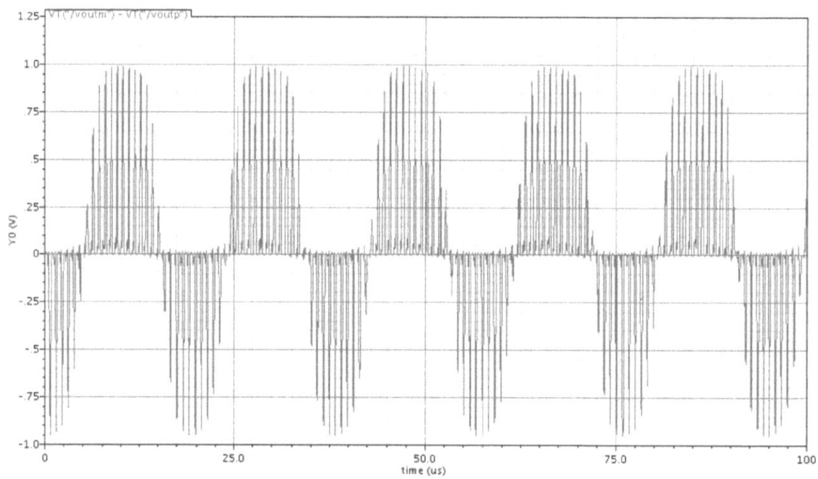


**Fig. 5.33** Simulated Switched-integrator differential output voltage for an input signal with a frequency value equal 100.05 MHz and amplitude value equal to 300 mV

The previous graph shows that the measured output voltage of the circuit should be a compressed sine wave shape. Since this is not the case, the problem must be located before the input terminals of the differential pair $M_1$. The best explanation is that the FIB connection between the metal 4 of the input pads and the metal 6 of the circuit is not perfect and the resulting circuit introduces the measured gain attenuation. This could happen because a parasitic capacitance between the two input terminals of the circuit was created by the FIB. This capacitance introduces attenuation in the signal path.

The gain of the circuit as a function of the input signal frequency is determined by repeating the previous procedure for different input signals with different frequency values and the same amplitude. The resulting transfer functions of the switched integrator circuit are depicted in Fig. 5.34.

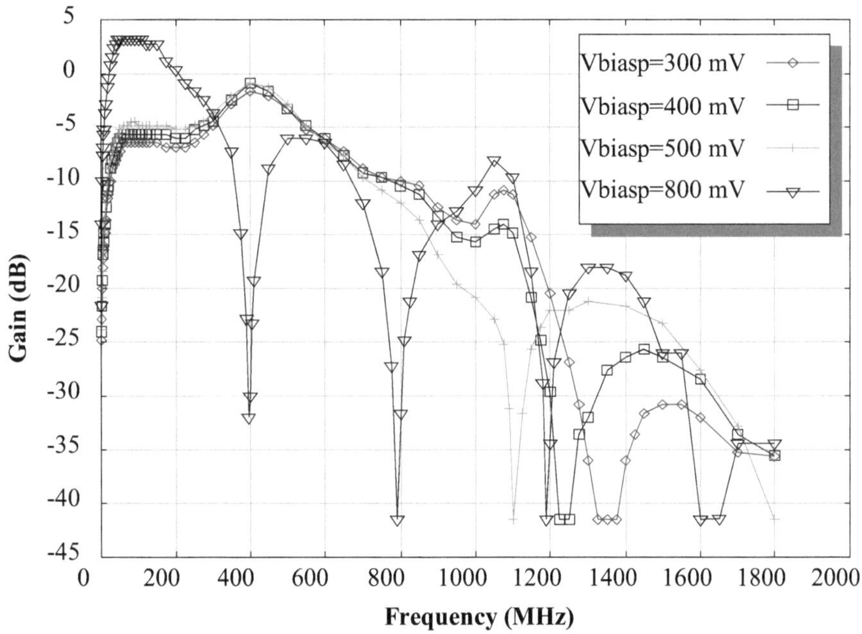


**Fig. 5.34** Measured Switched-integrator gain as a function of the frequency of the input signal, for four different bias voltage values

The previous graphs show that the transfer functions of the circuit are similar to a *sinc* function. The frequencies where the zeros appear are related to the values of the $V_{biasP}$ voltage, which controls the pulse width of the switched integrator. As expected the gain of the circuit increases with the $V_{biasP}$ voltage value (because the duration of the integration time interval increases) while the bandwidth decreases. From the previous graph it is possible to estimate the duration of the integration interval of the switched integrator, it is approximately equal to the inverse of the frequency value where the first zero occurs. It should also be noted that the experimental setup exhibits two resonant frequencies, one around 400 MHz and the other around 1100 MHz, these resonant frequencies are probably due to parasitic components in the PCB board. The high pass filter due to the input coupling capacitors has a cut off frequency value around 80 MHz.

The switched-integrator differential output noise voltage can be determined by calculating the standard deviation of the switched integrator differential output voltage, when the input voltage signal is disconnected. The termination resistance $R_t$ acts as the noise source of the input signal. Unfortunately, the noise level of the oscilloscope ADC is higher than the expected noise level of the switched-integrator circuit, resulting that the measured noise level would be essentially due to the oscilloscope. In order to measure the differential output noise of the circuit it is necessary to filter the noise created by the oscilloscope. This is done by averaging together several samples produced by the oscilloscope for the same output voltage sample of the circuit (see Appendix E). The switched integrator circuit uses the clock scheme depicted in Fig. 5.31; this means that there are 4 clock phases with a duration equal to 200 ns, each producing its own differential noise voltage value. If the oscilloscope ADC has a sampling frequency value equal to 1 GHz, the noise voltage of each clock phase will be sampled 200 times by the oscilloscope. If all these samples are added together the noise voltage of the ADC will be reduced to 1 mVrms (see Table E.1 in Appendix E), but the noise level from the circuit will remain unaffected because it is basically a DC signal, constant inside each 200 ns period. The averaging operation, as previously described in Chapter 3, corresponds to the *sync* transfer function; in this case the first zero is located at 5 MHz. The noise produced by the amplifier inside the switched integrator circuit is dominated by the flicker noise and will be essentially unaffected by the averaging operation. The digital oscilloscope is capable of recording sets of 10000 samples, therefore each set of data contains 50 periods with 200 ns of duration, each of these periods should correspond to a single noise voltage value produced by the circuit. In order to have enough data points to calculate the rms noise voltage of the circuit 12 sets of 10000 samples are recorded. An example of this procedure is shown in Fig. 5.35., where the original oscilloscope samples and the filtered signal are shown.

During each of the clock phases, the differential output noise voltage value is different. During two of these clock phases RESET=1, meaning that the output differential noise voltage is essentially due to the amplifier. When RESET=0 the output voltage is due to the amplifier noise and the reset switch noise stored in $C_2$ and after the rising edge of the SIG_T1 clock signal the noise from the switched integrator circuit is added. The average voltage values for each clock phase are grouped together and the corresponding standard deviation is calculated. The calculation of the standard deviation value uses only the last 160 samples in each 200 ns period in order to avoid the transient behavior of the circuit. The standard deviation value is due to the noise of the circuit and due to the filtered noise of the oscilloscope. In order to obtain the noise power of the circuit it is necessary to subtract the oscilloscope filtered noise power (approximately 1 mVrms) from the measured noise power. The measured rms output noise voltage values are shown in Table 5.7.

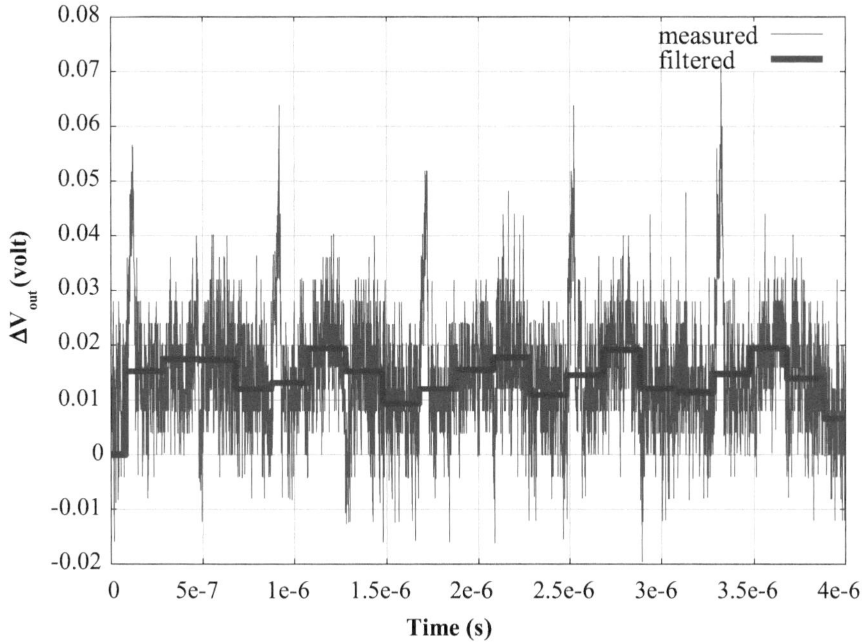


**Fig. 5.35** Measured Switched-integrator output differential voltage with no input signal for $V_{biasp}$=0.6 V and filtered output differential voltage

**Table 5.7** Measured rms output noise voltage values

| Clock phase | #1 | #2 | #3 | #4 |
|---|---|---|---|---|
| Measured rms voltage value (mVrms) | 3.31731 | 5.88313 | 3.52495 | 4.84908 |
| Circuit rms voltage value (mVrms) | 3.150 | 5.790 | 3.368 | 4.736 |

In the previous Table clock phases #1 and #2 correspond to RESET=0 and clock phases #3 and #4 correspond to RESET=1. The rising edge of the SIG_T1 clock signal occurs between the clock phases #1 and #2.The theoretical output noise voltage can be calculated using expression 5.3. Assuming an integrating interval of the switched integrator circuit equal to 1 ns (corresponding to the $V_{BiasP}$ = 0.6 V) and that the output noise voltage of the amplifier is 4 mVrms, results that the theoretical output noise voltage is equal to 4.43 mVrms. This value is fairly close with the measured noise voltage of the circuit. The input referred noise voltage of the circuit would be 23.5 dB higher because of the voltage gain problem.

### *5.2.3 Digitally Programmable Delay*

The VCDL is the block that is responsible for introducing a delay into a clock signal. The amount of delay is controlled by a voltage that in turn is determined by the delay locked loop. This voltage is generated internally by the differential to single ended conversion circuit (see Section 4.3.6 of Chapter 4) and is available through the pin $V_{Biasp}$ loop. By setting the bias current of this circuit to zero (this current is set through pin $I_B$ dsc) it is possible to impose externally any value for the control voltage. Therefore it is possible to measure the delay of the VCDL as a function of its control voltage; this is done by applying a clock signal with a frequency value equal to 2.5 MHz to the SIG_IN pin and measuring the delay between this clock signal and the output clock signal (pin SIG_T2). The resulting transfer function is shown in Fig. 5.36.

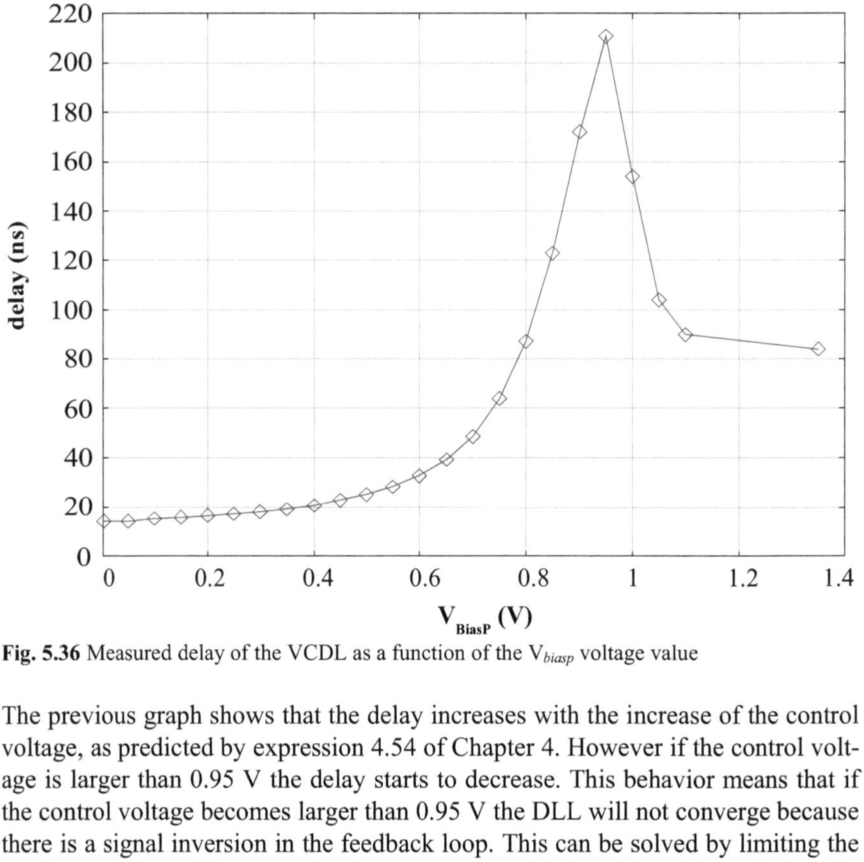


**Fig. 5.36** Measured delay of the VCDL as a function of the $V_{biasp}$ voltage value

The previous graph shows that the delay increases with the increase of the control voltage, as predicted by expression 4.54 of Chapter 4. However if the control voltage is larger than 0.95 V the delay starts to decrease. This behavior means that if the control voltage becomes larger than 0.95 V the DLL will not converge because there is a signal inversion in the feedback loop. This can be solved by limiting the maximum value of the control voltage to 0.9 V. This is achieved by reducing the bias current of the differential to single ended conversion circuit. The minimum delay value of the VCDL is equal to 14.3 ns. The duty-cycle of the output clock of the VCDL as a function of the control voltage is shown in Fig. 5.37.

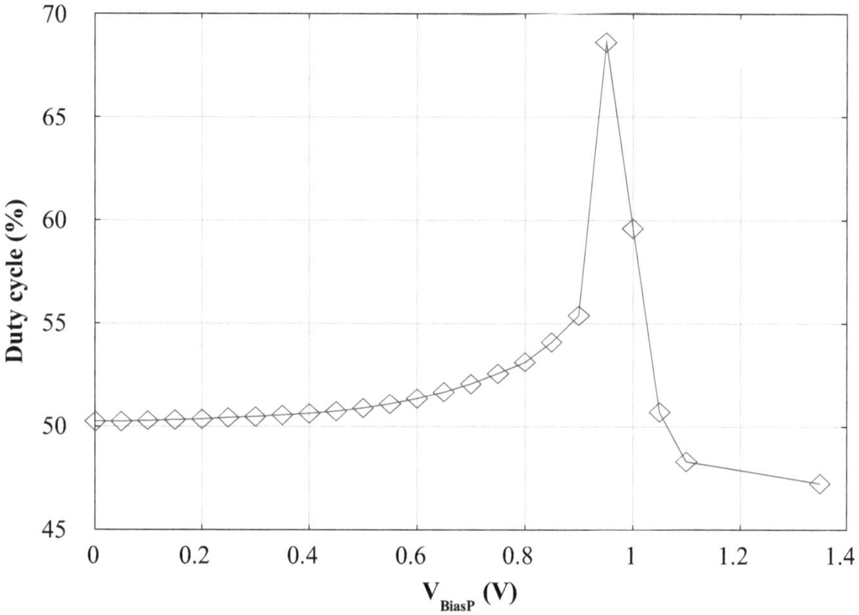


**Fig. 5.37** Measured duty-cycle of the output clock signal of the VCDL as a function of the $V_{biasp}$ voltage value

The duty-cycle of the output clock signal of the VCDL is inferior to 55% for delay values inferior to 100 ns. For larger delay values, the duty cycle of the output clock becomes very asymmetrical.

The digitally programmable delay uses a digital ΣΔ modulator to generate the reference clock of the DLL loop. Since this block is not included in the prototype IC, it is necessary to generate this clock signal. This is done using the digital pattern generator (Tektronix AWG 510) which is capable of creating 10 clocks signals with independent digital patterns. In order to program a certain amount of delay it is necessary to program the digital pattern generator to have a specific digital pattern describing the reference clock for that particular delay value. The digital patter is obtained using a script that simulates the behavior of digital ΣΔ modulator with the desired input value (delay) and based on the output bit stream creates the appropriate clock signals. The clock signals are written into an ASCII file that, in turn, is converted into a digital file (using a small software program from Tektronix) that can be loaded into the digital pattern generator. This procedure is used to create 10 files, each corresponding to a delay value multiple of 10 ns (10 ns, 20 ns, ..., 100 ns). As an example the clock signals for the case when the desired delay is 50 ns are shown in Fig. 5.38.

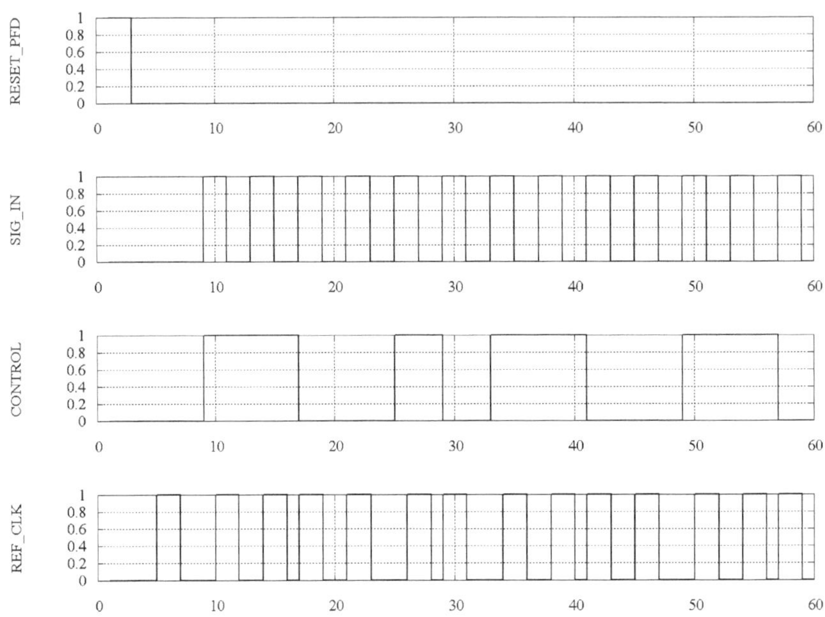


**Fig. 5.38** Clock signals used in the digitally programmable delay circuit (the CONTROL clock is the ΣΔ output and is not applied to the circuit)

The memory of the digital pattern generator is limited to 512 words, therefore the previous pattern is repeated after 126 clock cycles (each clock cycle uses 4 words, each word has a period equal to 100 ns). In the beginning of each pattern cycle the phase comparator is reseted to guaranty that it is in the correct state. For each of the 10 desired delay values, the actual delay between the input clock signal (SIG_IN pin) and the output clock signal (SIG_T2 pin) is measured using the oscilloscope. These measured results are shown in Fig. 5.39 and show that the circuit has the correct functionality. There is, however, a discrepancy between the programmed delay values and the actual measured delay values. The digital programmable delay circuit exhibits a non-linear programming characteristic. These non-linearities change with the power supply voltage and the bias current of the charge-pump circuit ($I_B$ CP) which seams to indicate that this circuit is not correctly designed and its output resistance is changing too much with the output voltage swing. However at this time it is not possible to guaranty that the problem resides solely in this circuit.

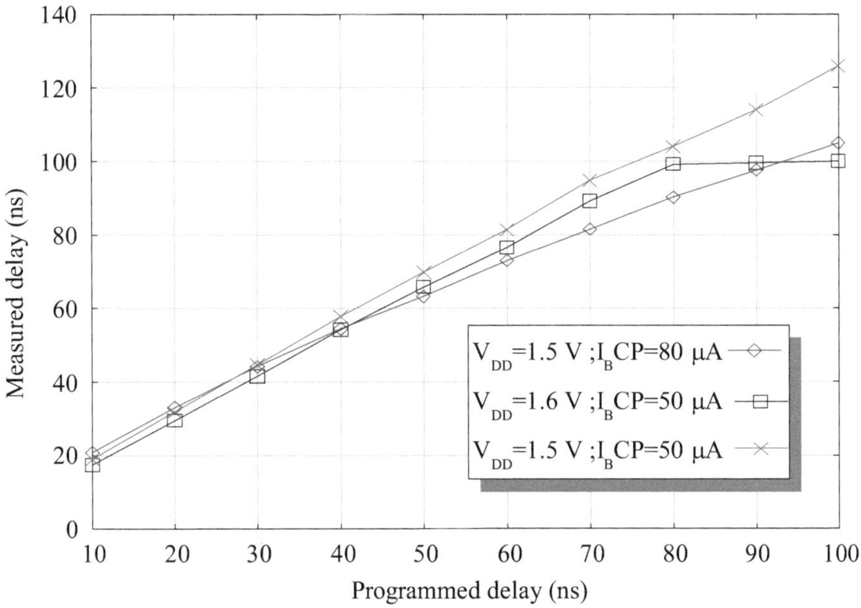


**Fig. 5.39** Measured delay as a function of the programmed delay value for different bias settings

# References

1. Grilo, J.; MacRobbie, E.; Halim, R.; Temes G.; A 1.8 V 94 dB dynamic range ΣΔ modulator for voice applications; 1996 IEEE International Solid-State Circuits Conference, 1996. Digest of Technical Papers. 43rd ISSCC, 8–10 Feb. 1996 Page(s):230–231, 451
2. Ribner, D.B.; Copeland, M.A.; Design techniques for cascoded CMOS op amps with improved PSRR and common-mode input range; IEEE Journal of Solid-State Circuits, Volume 19, Issue 6, Dec. 1984 Page(s):919–925
3. Alimenti, F.; Mezzanotte, P.; Roselli, L.; Sorrentino, R.; Modeling and characterization of the bonding-wire interconnection; IEEE Transactions on Microwave Theory and Techniques, Volume 49, Issue 1, Jan. 2001 Page(s):142–150
4. Chi-Taou T.; Package inductance characterization at high frequencies; IEEE Transactions on Components, Packaging, and Manufacturing Technology, Part B: Advanced Packaging, Volume 17, Issue 2, May 1994 Page(s):175–181

# Chapter 6
# Conclusions

**Abstract** this chapter describes an overview of the research presented in this book, draws some conclusions about obtained results and presents some suggestions for future work in this area.

The main objective of the work presented in this book was to study the feasibility of a low-cost low-power short range radar sensor. In order to achieve a radar sensor capable of differentiating between targets that are close together, it was necessary for the radar to transmit and detect pulses with a small time duration. If these pulses were to modulate a carrier signal (sine-wave), the resulting signal would have needed a carrier with a large frequency value. This would have prevented the use of CMOS technology. The solution was to use short duration pulses without carrier. These signals are known as UWB signals and are characterized by a large fractional bandwidth value. The large fractional bandwidth of these signals meant that the design techniques used for narrow band RF signals could not be applied. Basic concepts such as gain or noise figure had to be redefined, resulting in a new set of parameters used to evaluate the performance of the circuit. The wide bandwidth of the signals also means that the interaction between the circuit and the antenna is different from the case of narrow band signals.

The interaction between UWB signals and the antennas used to transmit and receive them was studied. In this study, a methodology that allows calculating the shape of the received UWB signal in an antenna link was developed. Using this methodology, it was show that the optimal input impedance of an UWB receiver is not necessarily the same as the optimal input impedance for a narrow band signal. The radar equation, usually defined for narrow band signals, was redefined for UWB signals. This new radar equation was used to analyze the echo signals from targets with basic shapes, resulting in an estimative of the echo signal amplitude as a function of the target distance. The architecture of an UWB radar was determined to be based in the concept of sub-sampling and the constituting blocks of this architecture were described. It was concluded that the sampling clock should be a delayed version of the clock signal used to transmit the radar pulses. The value of this delay controls the target range. The concept of SNR was redefined for the case of an UWB pulse and was used to evaluate the noise performance of the receiver channel of the radar. In order to improve this SNR, the S&H function was replaced by a switched-integrator function. A differential switched-integrator circuit was designed based on its noise performance. In order to improve the noise performance of the receiver channel, the output samples are added together. This could lead to an offset build-up in the output voltage, due to the mismatches between the components of the circuit. In order to reduce this effect offset calibration scheme was included in the circuit.



N. Paulino et al., *Low Power UWB CMOS Radar Sensors*, 211–213.

A new architecture for a digitally programmable delay that can have a large programming linearity was presented. This architecture is based on a digital ΣΔ modulator that controls a 1-bit digital to time converter, whose output is filtered by a delay locked-loop, thus producing the delayed clock signal. This architecture was analyzed and designed to meet the required specifications for the radar system. A high level model is used to simulate the behavior of this circuit and validate the design. The electronic sub-blocks necessary to build this circuit were described, analyzed and a design methodology, for each one, was derived. These circuits were implemented using differential clock signals in order to reduce the noise level in the radar system.

An experimental CMOS prototype IC was developed to evaluate the performance of the different sub-blocks of the radar system. A short-pulse generator circuit was developed, this circuit was used both to generate the transmitted pulse and the short pulse needed for the switched integrator circuit. The experimental evaluation procedure for the different sub-blocks was also described. The current consumption of this circuit was 5.4 mA. The experimental results show that the short-pulse generator is capable of creating UWB pulses with a time duration between 0.7 ns and 3 ns. The functionality of the switched-integrator circuit was verified. The frequency response of this circuit is as expected a *sinc* function that changes with the integrating pulse duration. However, the measured voltage gain of this circuit is 23.5 dB below the simulated voltage gain. This difference could be due to a layout error of the circuit that was later corrected using a FIB machine. The output noise voltage of the circuit was measured and the result is consistent with the design value. The functionality of the digital programmable delay was verified experimentally. The linearity of the programming resolution was smaller than expected and it was discovered that the measured delay value could be affected by the bias current of the charge pump circuit in the DLL circuit. This seems to indicate that this circuit was not correctly designed.

The experimental prototype demonstrated that the concept of implementing a low-power low-cost UWB radar in CMOS technology is feasible. The functionality of the constituting sub-blocks was demonstrated. However the performance of some of these blocks could be improved by a new design that could correct some of the problems found in some of the constituting sub-blocks. This design could use more area and more power in order to obtain a better noise performance. A complete radar transceiver could also be included if the area of the circuit is increased. There are some areas that were not explored in detail in this Thesis that could lead to future interesting results.

The transient response of an antenna link to UWB signals was calculated in Chapter 2; this was based on the inverse Fourier transform. This method required that the frequency response of the antenna is known. In this case, the frequency response was obtained by simulation. However the available simulating software was only capable of analyzing antennas that are constituted by wires. This means that several types of antennas were not analyzed. It is important to learn more about the interaction of the antenna with the UWB signals, since this could result in better designs of antennas for these types of signals. The antenna could be designed at the same time as the circuit that is connected to. This could result in a better wide-band impedance matching, than designing the circuit and the antenna separately.

In this work it was decided not to use a LNA. However, in certain cases these circuits must be used. From the literature about wide band LNA circuits it is possible to conclude that there is a long way to go before all the issues related to wide-band LNA are solved. The main difficulties are the large noise power in these circuits and the wide-band impedance matching required for the input.

In this Thesis, it was assumed that the target would be detected by determining when the echo signal amplitude is larger than given reference level. More advanced algorithms and signal processing techniques should be used to improve the target detection and identification. Several receivers can be combined in order to obtain a 2-D or 3-D representation of the target.

# Appendix A
# Echo Signals from Moving Targets

If a target is moving, the echo signal it emits is different from the one created by the same target when it is not moving. In order to understand the effect of a moving target in the echo signal, the significant events occurring when a moving target is detected by a radar system will be analyzed. In Fig. A.1, the simplified diagram of a radar and moving target is shown. At time $t$=0 the radar is starting to emit a pulse and the target, located at a distance $d$, is moving towards the radar with a speed $v_t$ which is considered to be much smaller than $c$.

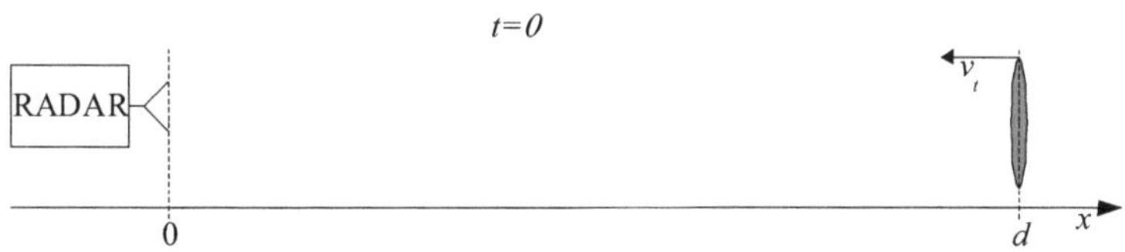


**Fig. A.1** Simplified diagram of a radar detecting a moving target

At time instant τ the radar finishes emitting the pulse and the target moved $c \cdot \tau$ toward the radar. The pulse travels at the speed of light $c$ towards the target (Fig. A.2).

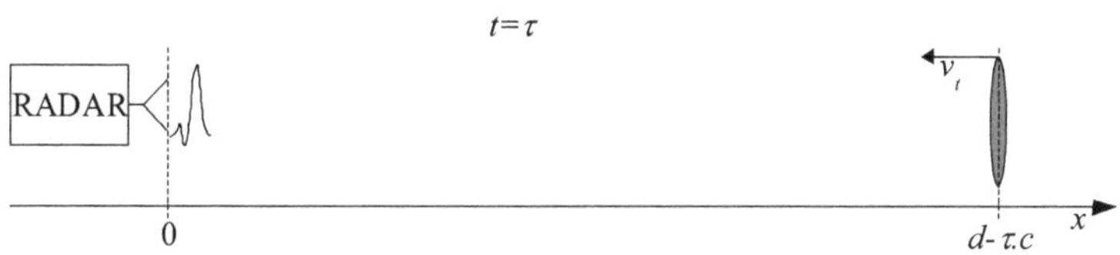


**Fig. A.2** The radar emits a pulse with duration equal to τ, during this time the target moves a distance $\tau.c$ towards the radar

At time instant $T_1$, after traveling a distance equal to $c \cdot T_1$ the leading edge of the pulse hits the target and starts to generate the echo signal (Fig. A.3). The target position at this time instant is $d_1 = d - T_1 \cdot v_1$. The time instant $T_1$ can be calculated using:

$$d - T_1 \cdot v_t = c \cdot T_1 \Leftrightarrow T_1 = \frac{d}{v_t + c} \tag{A.1}$$

The position at time instant $T_1$ is equal to

$$d_1 = d - T_1 \cdot v_t = \frac{d \cdot c}{v_t + c} \tag{A.2}$$

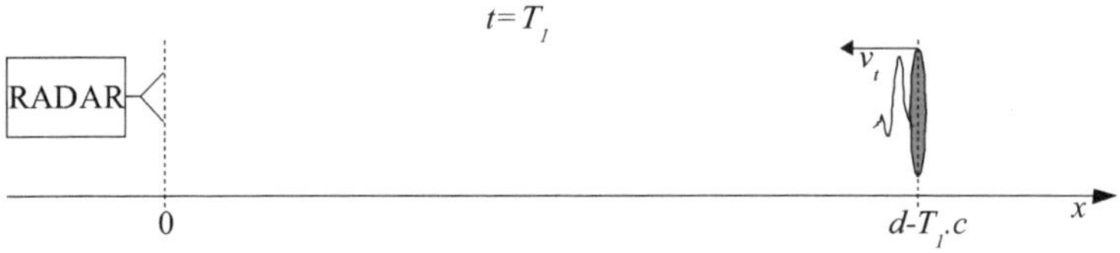


**Fig. A.3** The leading edge of the pulse hits the target at time instant $T_1$

At time instant $T_2$, the pulse passes past the target and the echo signal is now completely formed and traveling back to the radar. This is shown in Fig. A.4.

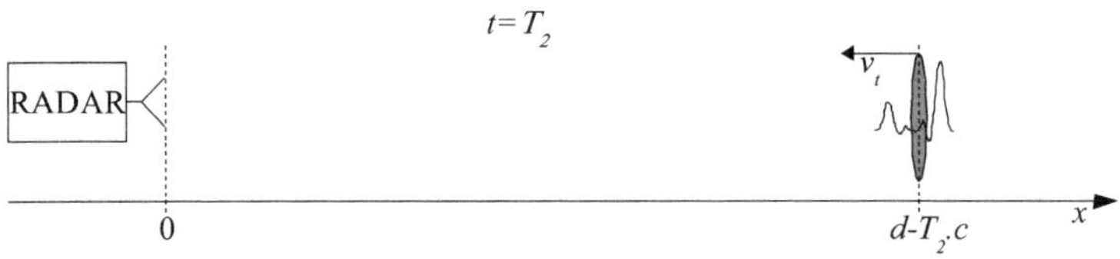


**Fig. A.4** The pulse passes past the target and the echo is fully formed and is traveling back to the radar

Assuming that the target does not introduce dispersion into the echo signal, the time instant $T_2$ can be calculated using (note that the trailing edge of the pulse was radiated at time instant $\tau$):

$$d - T_2 \cdot v_t = c \cdot (T_2 - \tau) \Leftrightarrow T_2 = \frac{d + \tau \cdot c}{v_t + c} \tag{A.3}$$

The width of the echo signal is calculated using expressions A.1 and A.3, and it is equal to:

$$\tau_{echo} = T_2 - T_1 = \tau \cdot \frac{c}{v_t + c} \tag{A.4}$$

Expression A.4 shows that the returning echo signal is shorter than the radiated pulse width. This is because the target is moving towards the radar, if it was moving away from the radar the echo pulse would be longer than the transmitted pulse. If the target speed is small, this effect can be negligible. For example, if the target is moving with a speed equal to 100 m/s (360 km/h) the pulse distortion (compression or expansion) would be smaller than 0.0004%, which is clearly very difficult to measure.

The leading edge of the echo signal arrives at the radar at time instant $T_2$, as show in Fig. A.5.

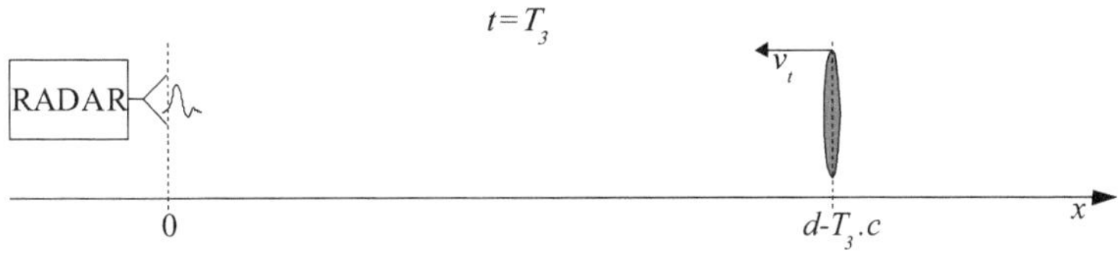


**Fig. A.5** The echo signal arrives at the radar

The time instant $T_3$ can be calculated using:

$$d_1 = c \cdot (T_3 - T_1) \Leftrightarrow T_3 = \frac{2 \times d}{v_t + c} \quad \text{(A.5)}$$

The radar receiver determines the target range using the expression 3.1, resulting that the target range computed by the radar is equal to:

$$R = \frac{c \cdot T}{2} = \frac{c}{2} \cdot (T_3 - 0) = \frac{d \cdot c}{v_t + c} \quad \text{(A.6)}$$

which is the distance from the radar where the target was located when the leading edge of the pulse hit it (see expression A.2).

In the case when the radar is continually ranging the target, the radar emits pulses with a frequency equal to *PRF* (Pulse Repetition Frequency), continually creating echo signals. Since the target is moving towards the radar its distance to the radar is reducing, hence each echo will arrive early compared to the previous one. Considering that the first pulse is emitted at time instant 0, the arrival time of the first echo occurs at time instant $T_3$ (given by expression A.5). The next pulse is emitted at time instant $T_{PRF}$ (one PRF period after the first pulse), during the time between pulses the target has moved to $d - T_1 \cdot T_{PRF}$. Replacing the original target distance (d) by this new distance in expression A.5 and adding $T_{PRF}$, results in the arrival of the second pulse equal to:

$$T_4 = 2 \times \frac{d - v_t \cdot T_{PRF}}{v_t + c} + T_{PRF} \quad \text{(A.7)}$$

The period of the echo signal is given:

$$T_4 - T_3 = 2 \times \frac{d - v_t \cdot T_{PRF}}{v_t + c} + T_{PRF} - 2 \times \frac{d}{v_t + c} = T_{PRF} \cdot \left( \frac{c - v_t}{v_t + c} \right) \quad \text{(A.8)}$$

Note that the period of echo signal is smaller than the period of the transmitted pulse ($T_{PRF}$) because the target is moving toward the radar. This is know as the Doppler effect. If the target was moving away from the radar, it is necessary to introduce a negative target velocity value in the expression, resulting in a higher period value.

If the target speed value is small the difference between the transmitted pulse period and the received echo period is very small. For example a *PRF* value of 1 MHz and a target velocity of 100 m/s results in a difference less than 0.7 Hz between the *PRF* and the echo signal frequency.

A short range radar system (with a maximum range inferior to 20 m) must deal with targets moving at slow speeds, or otherwise the target would quickly move out of the radar range. In such a system a target moving faster than 1 m/s would only be detectable for less than 20 s. Therefore a maximum target speed considered is 1 m/s. A a target with this speed would move 1 μm between each transmitted pulse for a *PRF* value equal to 1 MHz, if the PRF value is lowered to 1 kHz, the target would move 1 mm, which is still well below the radar resolution.

# Appendix B
# Gain and Offset Errors in the S&H Due to the MOS Switches

A MOS transistor is not an ideal switch, therefore it is important to understand how the non-ideal behavior of the transistor affects the performance of a switched-capacitor S&H circuit. An ideal switch has a zero ON resistance and an infinite OFF resistance. A MOS transistor, used as switch, exhibits a very large OFF resistance (with a very low leakage current), but has a finite ON resistance that is inversely proportional to the *W*/*L* ratio of the transistor. This means that in order to have an acceptable low value for the ON resistance the transistor the channel length (*L*) should be as small as the technology allows and the channel width (*W*) should be larger than a given value in order to obtain an ON resistance below the desired value. Another issue associated with the switch ON resistance is that these resistance is non-linear, i. e. its value depends on the applied voltage value ($V_{DS}$). This is not a problem in the S&H circuit since the input signals have a very small amplitude and therefore the amount of distortion introduced by the non-linear switch resistance is very small.

A small ON resistance can obtained by increasing the width of the transistor, however, this also increases the transistor parasitic capacitance $C_{DB}$ and $C_{SB}$., as it is depicted in Fig. B.1

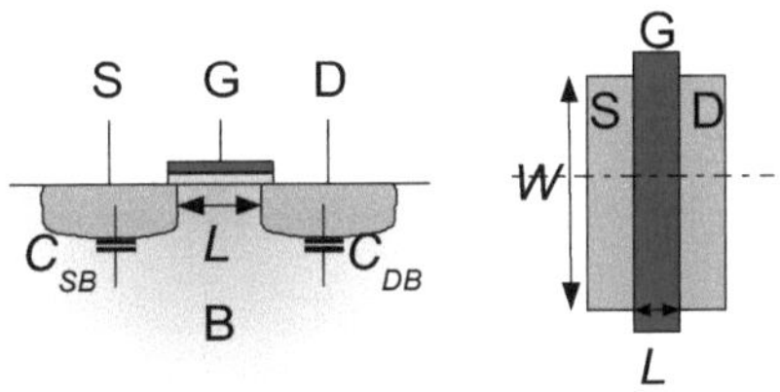


**Fig. B.1** MOS transistor physical layout. On the right side is the top view of a MOS transistor and on the left side is the cut view of the same transistor. This diagram shows that increasing the channel width (*W*) increases the parasitic capacitances $C_{DB}$ and $C_{SB}$. The parasitic capacitances are due to the inversion layer between the diffusion areas of the source (S) and drain (D) of the transistor and the substract (B)

The MOS parasitic capacitances cause the settling error in the S&H circuit to be larger because the *RC* time constant must now include these capacitances and it is larger. This means, that in order to determine the acceptable value of the transistor width (*W*) the switch, it is necessary to evaluate the impact of the transistor parasitic capacitances in the settling error.

Considering a differential S&H circuit, such as the one depicted in Fig. B.2, it is possible to simulate the circuit and obtain the value of the sampled voltage in capacitors $C_1$ and $C_2$ at the end of the sampling phase.

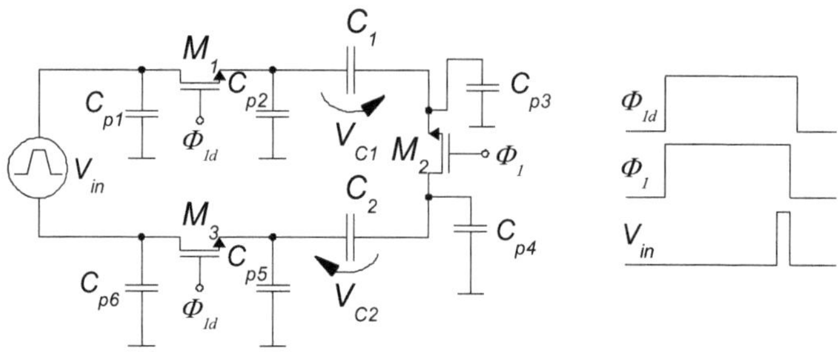


**Fig. B.2** Differential S&H switched-capacitor circuit, with transistor parasitic capacitances to ground. The circuit uses a delayed phase scheme (shown in the right side of the figure) to reduce the signal dependent charge injection

The circuit was simulated using transistors with minimum channel length and increasing channel width (all transistors have the same size). For each channel width, the sampled voltage was measured at the end of the sampled phase. The input signal was a rectangular pulse with 1 ns duration and 10 mV amplitude value. The resulting sampling voltage ($V_{C1}+V_{C2}$), for $C_1 = C_2 = 7$ pF, is shown in the graph of Fig. B.3.

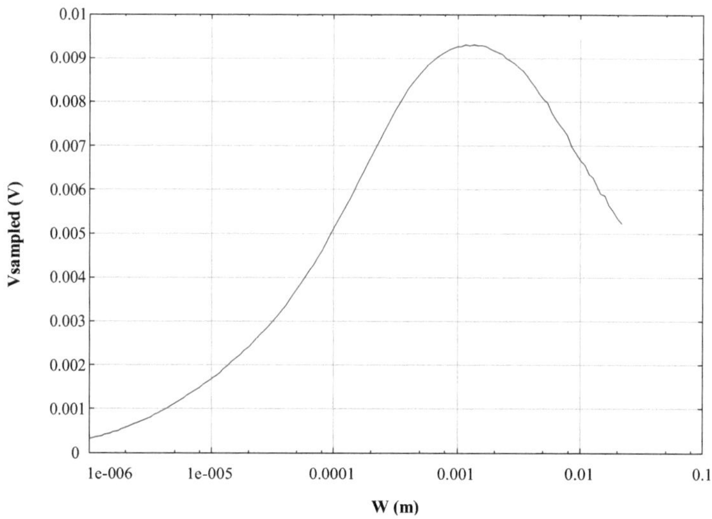


**Fig. B.3** Differential sampled voltage at the end of the sampling phase for varying channel widths in the switch transistors

The previous graph shows that the sampled voltage value increases when the channel width ($W$) of the transistors is increased, as expected, however after a certain value of $W$ the value of the sampled voltage starts to decrease. This is because after that value of $W$, the increase in the parasitic capacitances of the transistors is larger than the decrease in the transistor ON resistance. The graph shows that this occurs for a channel width approximately equal to 1 mm and for this value of $W$ the maximum sampled voltage is 9 mV, which is only 90% of the input pulse amplitude. For this value of $W$ the transistors have an ON resistance value around to 3 Ω (resulting in a total resistance around 9 W) and a parasitic capacitance around to 2 pF (resulting in a total parasitic capacitance value around 12 pF). In order to reduce this value it is necessary to use a better CMOS technology, either with a smaller minimum channel length ($L_{min}$) value (in this example a 0.35 μm technology was used) or with smaller parasitic capacitance values or with a larger $K_P$ value. Usually, all these last two factors are improved when a technology with a smaller $L_{min}$ value is used.

Using MOS transistors as switches also has other non-ideal effects: clock feed-through and charge injection. The first is due to the $C_{GD}$ and $C_{GS}$ parasitic capacitances that provide a path to the clock signal into the sampling capacitors, this effect is increased when the ratio between the sampling capacitances and the parasitic capacitances decreases. The second is due to the charge necessary to create channel in the transistor, when the transistor is turned ON the charge comes from the source and drain diffusion areas of the transistor and therefore these node experience a voltage change, but since the sampling capacitances are being charged by the input voltage signal this is not important. When the transistor is turned OFF the channel present in the channel can not be destroyed and it is absorbed by the source and drain nodes of the transistor. Once again, this will cause the voltage on these nodes to change, but since there is not input signal to charge the sampling capacitances, this translates into a sampling error voltage. The amount of charge present in the transistor channel is approximately constant, but it can change with the input signal voltage value. This dependency of the channel charge on the input signal causes a distortion of the sampled signal and it is necessary to minimize its effects. Usually this is achieved using differential circuits and a delayed phase scheme (bottom plate sampling), such as the one depicted in Fig. B.2.

The clock feed-trough from transistor $M_2$ causes a $\Delta V$ voltage step in both the drain and source nodes of the transistor, therefore the sampled differential voltage ($V_{C1}+V_{C2}$) is not affect by the clock feed-trough due to transistor $M_2$.

$$V_{C1}+V_{C2}=\left(V_{CP2}-V_{CP3}-\Delta V\right)+\left(V_{CP4}-V_{CP5}+\Delta V\right)=V_{CP2}-V_{CP5} \tag{B.1}$$

Assuming that transistors $M_1$ and $M_3$ have exactly the same size they will case cause the same $\Delta V$ voltage step in the nodes of $C_{p2}$ and $C_{p5}$, once again the sampled differential voltage value is not affected by $\Delta V$ because the two contributions are canceled. Of course, in reality, transistors $M_1$ and $M_3$ are not exactly equal; a small signal independent error (offset) is introduced into the sampled voltage due to the difference between $\Delta V_1$ and $\Delta V_3$. But, because the switch $M_2$ is opened before switches $M_1$ and $M_3$ these voltage steps are applied to $C_1$ in series with $C_{P3}$ and $C_2$ in series with $C_{P4}$. This results in a voltage division across these capacitors and since in the general case $C_1$ and $C_2$ are larger than $C_{P3}$ and $C_{P4}$ the offset will be very small.

In the falling edge of $\Phi_1$ the charge present in the channel of transistor $M_2$ is equally split between its drain and source nodes causing a $\Delta V$ voltage step in both these nodes. Therefore the sampled differential voltage is not affect by the charge injection of transistor $M_2$. The charge injection of transistors $M_1$ and $M_3$ causes a similar problem to the clock feed trough problem, with the difference that the channel charge is dependent on the input voltage and thus the error voltage is also signal dependent. This error voltage value is dependent of the mismatch between transistors $M_1$ and $M_3$ and the values the ratio between the values of $C_{p3}$ and $C_1$ or ($C_{p4}$ and $C_2$). If the input signal amplitude is small, the error voltage due to the charge injection can be considered as an offset signal.

To determine the value of the error voltage, the same circuit was simulated with an input signal of 0 V amplitude and the differential sampled voltage was measured at the end of the sampling phase. The channel width of transistor $M_1$ was set to be 1% larger than the channel width of the other transistors. This simulation was repeated for different channel width ($W$) (maintaining the same relative mismatch between the transistors). The resulting sampling voltage ($V_{C1}+V_{C2}$) is shown in the graph of Fig. B.4.

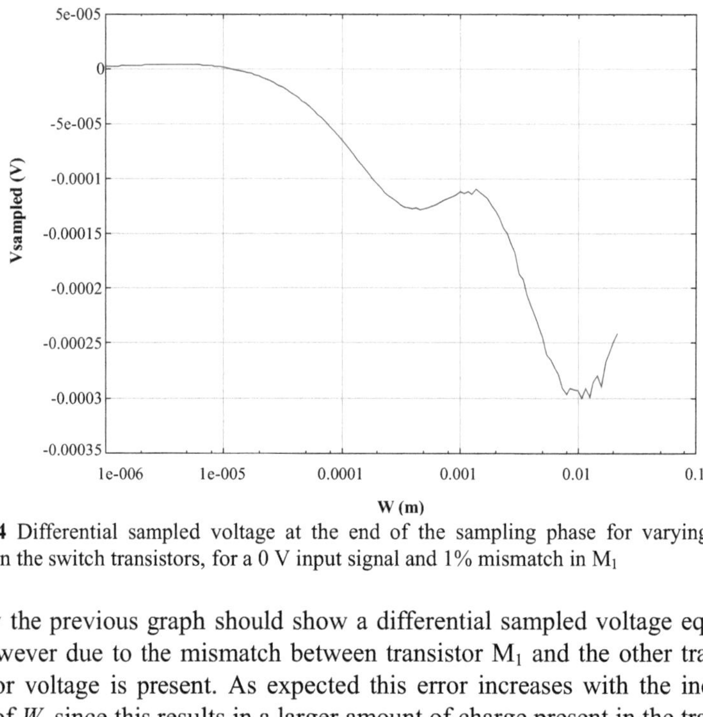


**Fig. B.4** Differential sampled voltage at the end of the sampling phase for varying channel widths in the switch transistors, for a 0 V input signal and 1% mismatch in $M_1$

Ideally the previous graph should show a differential sampled voltage equal to 0 V. However due to the mismatch between transistor $M_1$ and the other transistors an error voltage is present. As expected this error increases with the increasing value of $W$, since this results in a larger amount of charge present in the transistors channel and larger parasitic capacitances in the transistors and the value of the sampling capacitances is kept constant.

This graph shows that using transistors with a channel width of 1 mm results in an error voltage larger than 100 μV.

# Appendix C
# Sigma Delta Modulator

In this appendix the sigma delta modulator behavior is analyzed and described. The block diagram of a second order sigma delta modulator is shown in Fig. C.1. It is constituted by two discrete time integrators followed by a 1-bit quantizer, the output of the quantizer is feedback to the integrator inputs. The output of the 1-bit quantizer is equal to 1 if $y_2>0$ and is equal to –1 if $y_2<0$ (1 is represented by the logic level "1" and –1 by the logic level "0"). The signals in the sigma delta are sampled at a frequency equal to $F_{CLK}$.

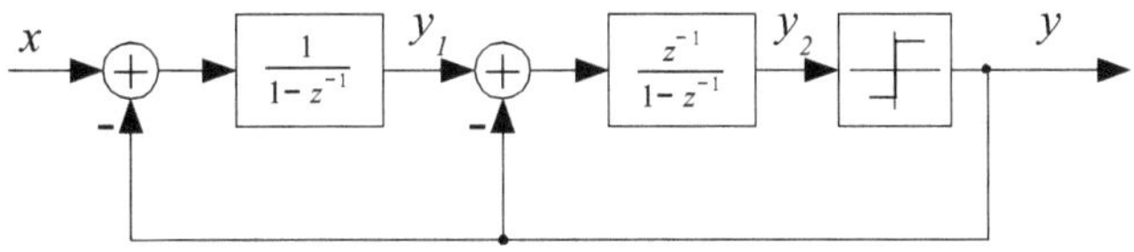


Fig. C.1 Block diagram of a second order ΣΔ modulator

Assuming that the signal $y_2$ changes randomly, it can be considered that the error introduced by the quantizer is uncorrelated to the input signal ($x$) and therefore is considered a separate noise signal ($N_Q$). Using this approach the 1-bit quantizer can be replaced by a gain equal to one. The output signal ($y$) is obtained adding the quantization noise ($N_Q$) to the signal $y_2$. There are two transfer functions that can be defined for this circuit, the signal transfer function (*STF*) and the noise transfer function (*NTF*). These are calculated next.
The output signal is given by:

$$y = N_Q + \frac{z^{-1}}{1-z^{-1}} \cdot \frac{1}{1-z^{-1}} \cdot (x-y) - y \cdot \frac{z^{-1}}{1-z^{-1}} \tag{C.1}$$

Solving this equation for $y$ results in :

$$y = N_Q \cdot \left(1-z^{-1}\right)^2 + x \cdot z^{-1} \tag{C.2}$$

The two transfer functions are given by:

$$\begin{aligned} NTF(z) &= \frac{y}{N_Q} = \left(1-z^{-1}\right)^2 \\ STF(z) &= \frac{y}{x} = z^{-1} \end{aligned} \tag{C.3}$$

From these transfer functions it is clear that the input signal is unaffected by the modulator (it is only delayed one clock period) and that the quantization noise is attenuated at low frequencies since the *NFZ* has a double zero at DC, this effect is referred to as noise shaping. The *NTF* transfer function is an approximation to the real modulator behavior. The modulator contains a highly non-linear element in its loop (the 1-bit quantizer) which introduces large errors into any linear analysis of the sigma delta modulator. To verify the performance of the sigma delta it is necessary to simulate the block diagram depicted above. The simulation is done calculating the values of the loop variables $y_1$, $y_2$ and $y$ at discrete time instants ($n$) using the following equations:

$$\begin{aligned} y[n] &= Sign(y_2[n-1]) \\ y_1[n] &= x + y_1[n-1] - y[n] \\ y_2[n] &= y_1[n] + y_2[n-1] - y[n] \end{aligned} \quad \text{(C.4)}$$

Applying an input signal equal to 0.5 results in the following output signal (Fig. C.2):

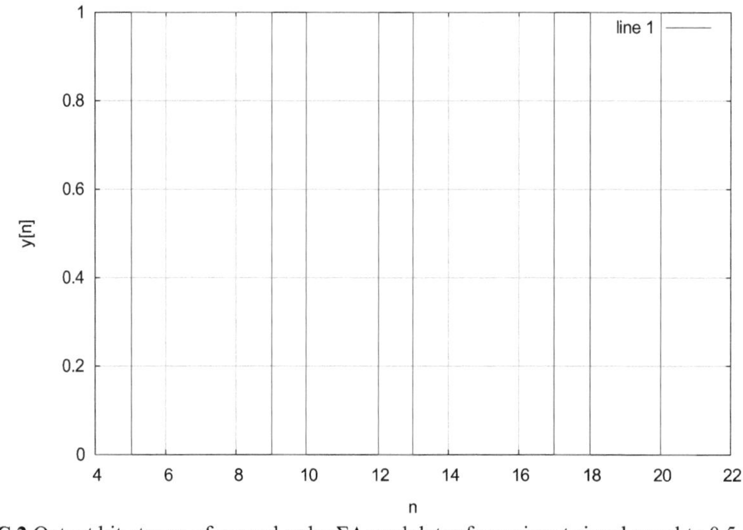


**Fig. C.2** Output bit-stream of second order ΣΔ modulator for an input signal equal to 0.5

This signal is composed by the repeating sequence: "01111011", the average value of the sequence is 6/8=0.75, since the output levels of the quantizer are –1 and 1, the equivalent output value is $0.75\times 2-1=0.5$ as expected. Note that if the input signal is equal to 1 then the output signal would be equal to "1111..." (an infinite sequence with an average value equal to 1), if the input signal is set to 0, the output signal would be the sequence "101010...". The spectral power density (spectrum) of the output signal can be estimated using the FFT and compared to the expected output spectrum (predicted by the *NTF*), this is shown in Fig. C.3:

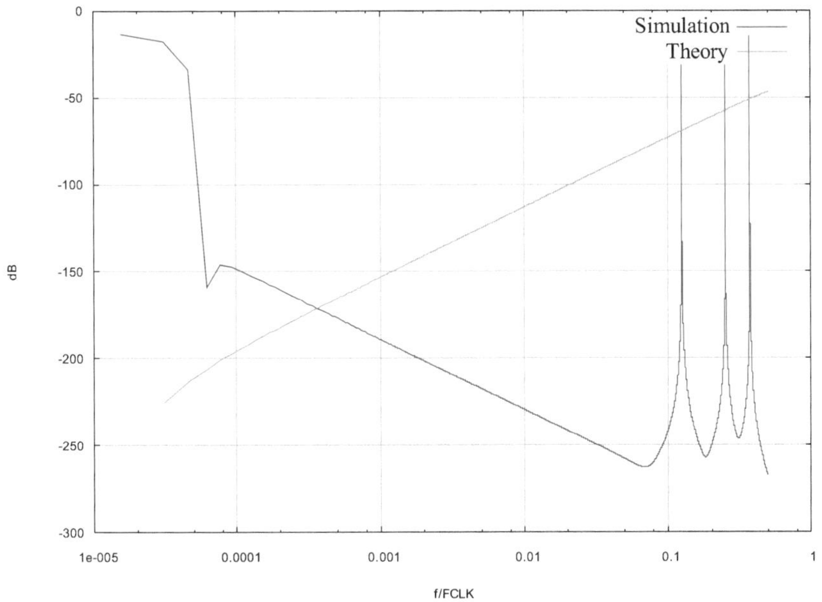


**Fig. C.3** Estimation of the PSD of the output bit-stream of second order ΣΔ modulator for an input signal equal to 0.5

In this graph it is clear that the energy of the quantization noise is concentrated into 3 frequencies ($0.5\times F_{CLK}, 0.25\times F_{CLK}, 0.125\times F_{CLK}$). These frequencies can be explained considering the output signal average value as 0.75=0.5+0.25, to produce an average value of 0.5 the minimum sequence length is 2 and to produce 0.25 the minimum sequence length is 4 combining these two sequences results in a sequence with a length of equal to 8. If the input signal value is set to 0.511 the output signal does has a very long repeating sequence, as shown in Figs. C.4 and C.5:

This sequence has a spectrum that approaches better the predicted spectrum of the quantization noise of the sigma delta modulator as it shown below:

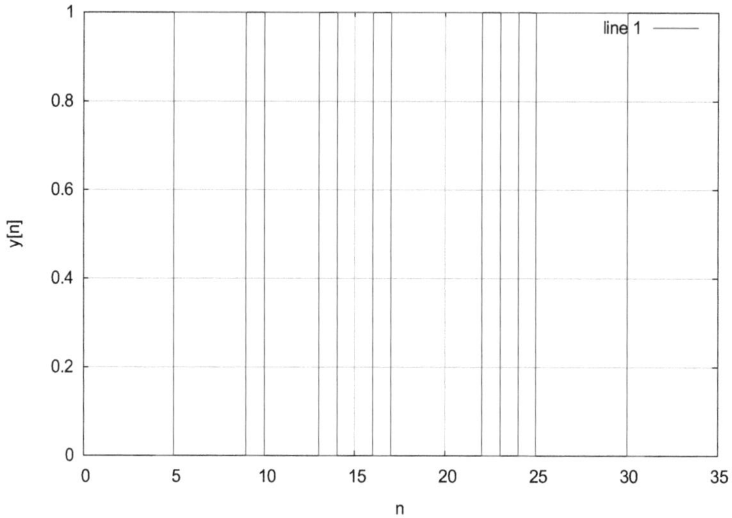


**Fig. C.4** Output bit-stream of second order ΣΔ modulator for an input signal equal to 0.511

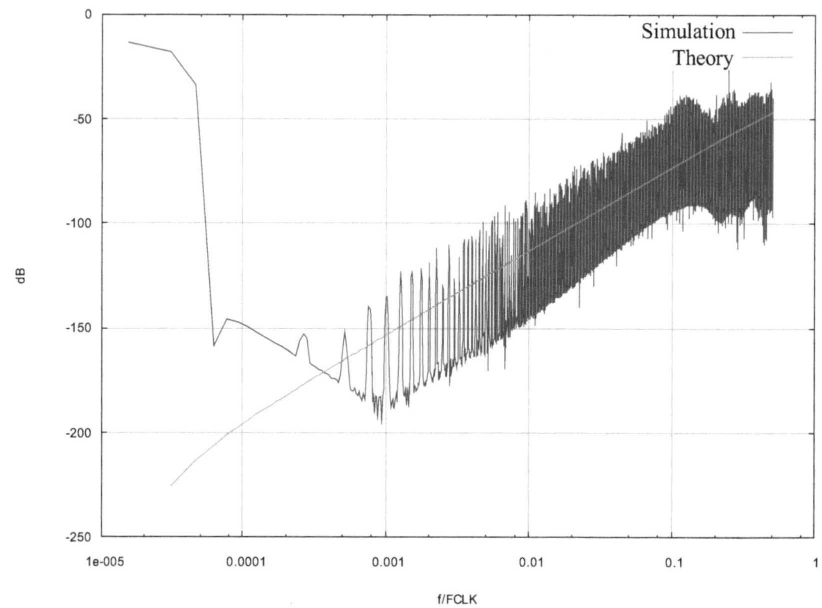


**Fig. C.5** Estimation of the PSD of the output bit-stream of second order ΣΔ modulator for an input signal equal to 0.511

The output spectrum is closer to the predicted quantization noise spectrum, but the power of the quantization error is concentrated into a multitude a frequencies, instead of being uniformly distributed across the frequency. This is a consequence that the output signal is a pattern of "1's" and "0's" with a very long repetition period (this pattern has several sub-sequences, each accounting for the multiple tones present in the spectrum). In order to spread the quantization power more uniformly across the frequency a dither signal should be added to the input. This is a signal with a small amplitude and high frequency. The objective of this signal is to break the cycle that the modular settles into, when the input is a DC value. The ideal dither signal is a random signal, but a single tone at a high frequency can also work (and is much simpler to generate). The following spectrum (Fig. C.6) is obtained for an input signal equal to 0.511 plus a dither signal equal to $(0, 2^{-10}, 0, -2^{-10}, \ldots)$.

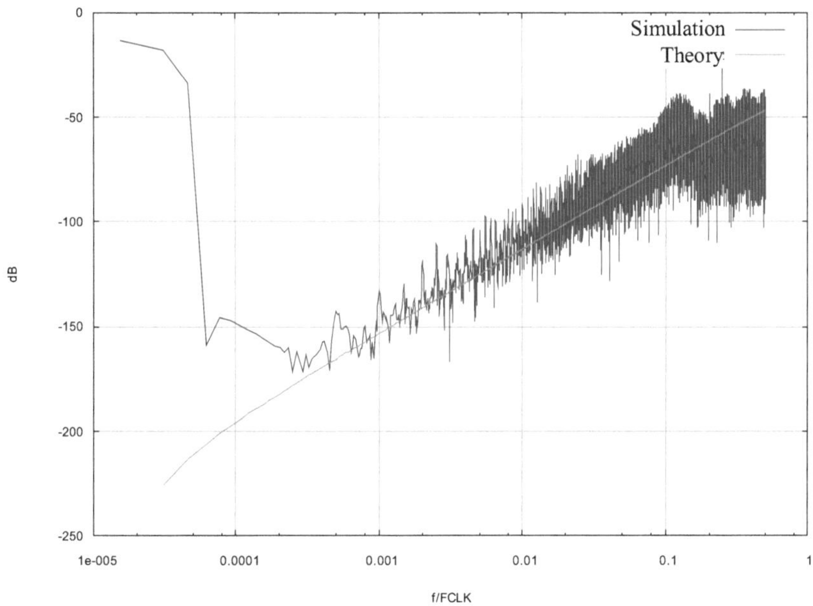


**Fig. C.6** Estimation of the PSD of the output bit-stream of second order ΣΔ modulator for an input signal equal to 0.511 plus a dither signal.

The spectrum of the output signal is closer to the theoretical spectrum of the quantization noise. There are still some tones in the spectrum but most of the quantization noise power is now uniformly distributed.

The objective of using the sigma delta modulator is to "push" the quantization noise power from the lower frequencies to the higher frequencies. The modulator is followed by a low pass filter that eliminates most of the quantization noise power (which is essentially at high frequencies). The resulting in band quantization noise power can be determined by integrating power spectral density of the output signal of the modulator up to the low-pass filter cut-off frequency. The in-band quantization noise power changes with the value of the input signal, since the modulator is non-linear system. The exact value of the quantization noise power after the low-pass filter is very difficult to predict because depending on the value of the input DC signal the output pattern can have more or less tones inside the filter band. This is well illustrated comparing the spectrum of the output signals for the 0.5 and 0.511 inputs. In the first case there are only 3 tones at high frequency and the quantization noise power would be zero inside a band up to $0.1\times F_{CLK}$. In the second case there are tones down to 0Hz, which results in a non-zero value for the quantization noise power after the low pass filter (the exact value would change with the cut-off frequency of the filter).

The modulator was simulated for DC input levels from –1 to 1 and the quantization noise power inside a band up to $F_{CLK}/64$ was calculated, these obtained noise power values are shown in the graph of Fig. C.7.

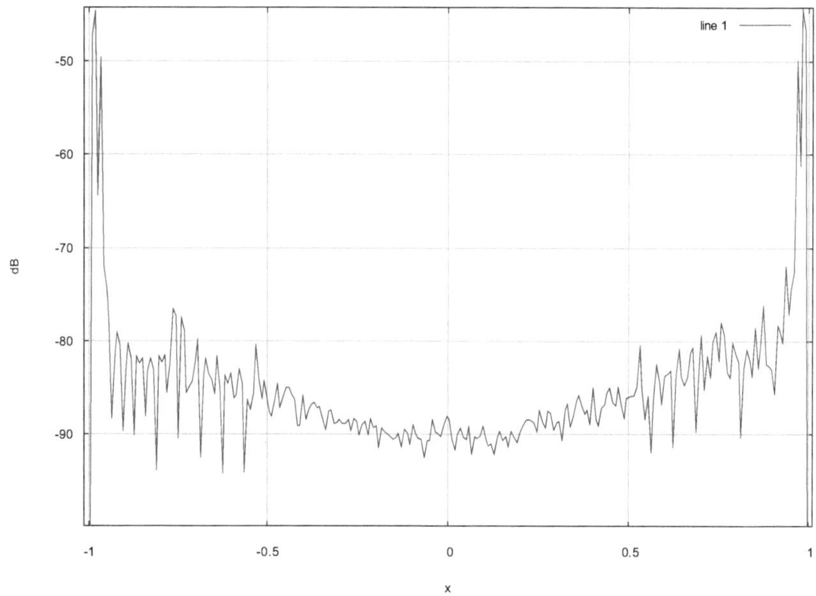


**Fig. C.7** Quantization noise power inside the band, as a function of the DC input value

This graph shows that the quantization noise power inside the band depends on the input signal value. The value of the noise power inside the band can have a large increase or decrease for a small variation of the input signal value. This is because depending on the input signal value a tone can appear or disappear from the band. The quantization noise power increases, on average, as the input signal approaches the 1 or the –1 values. If the input signal is close to these values, the output signal pattern will have long sequences of “1's” and “0's”, so that its average value can be close to 1 or –1. A longer output pattern is more likely to have long repeating sub-sequences, which result in tones inside the band and therefore result in an increase of the quantization noise inside the band.

# Appendix D
# Bond-Wire Parasitic Effects

In this appendix, the parasitic effects due the bond-wire connection between the silicon die and the PCB are analyzed. This analysis is used to determine the attenuation and the distortion created by the bond-wire connection.

The circuit in the silicon die is connected to the PCB or to the package using a thin wire (typically with 25 μm of diameter). This wire is connected to pads in the circuit, which are protected from ESD phenomenon by diodes. These protection devices, together with the large metal square that forms the pad itself, introduce a parasitic capacitance to ground. The value of this parasitic capacitance ($C_{pad}$) is typically smaller than 1 pF. On the other side the bond-wire connects to a package pin or to the PCB directly, this connection has also a parasitic capacitance to ground ($C_{pin}$) that typically is larger than the parasitic pad capacitance. If a package is used there are two parasitic capacitances, one due to the PCB trace and the other due to the package pin. The bond-wire can be considered as an inductance ($L_{bond}$) in series with a resistance ($L_{bond}$). Typically the parasitic resistance value is smaller than 1 Ω, the inductance of the bond-wire is typically 1 nH/mm. Therefore connecting a voltage source trough a bond-wire connection results in introducing a parasitic circuit, shown in Fig. D.1, in the signal path.

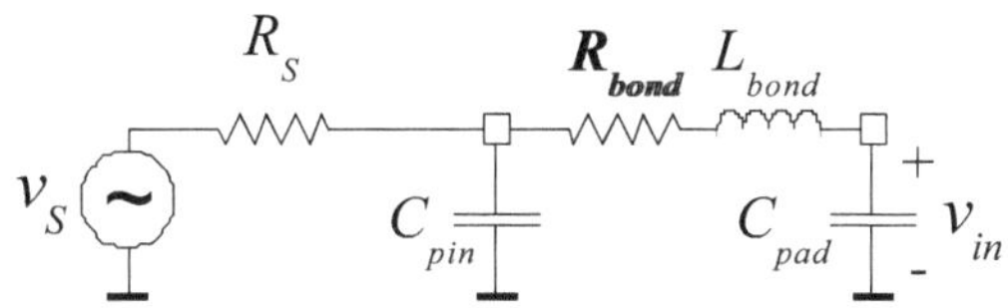


**Fig. D.1** Electrical circuit formed by the parasitic components of the bond-wire connection

The previous circuit is a third order low pass-filter. This circuit will filter the input signal and change its PSD. In order to verify this effect, the frequency response of the circuit was calculated. The voltage gain of the circuit, defined as the ratio between the voltage $v_{in}$ and the voltage $v_s$ is calculated as a function of the frequency. This was done for 3 different values of the bond-wire inductance and for $C_{pad}$ =2 pF, $C_{pin}$ =4 pF and $R_{bond}$ =0.1 Ω, the resulting voltage gain graph is shown next in Fig. D.2.

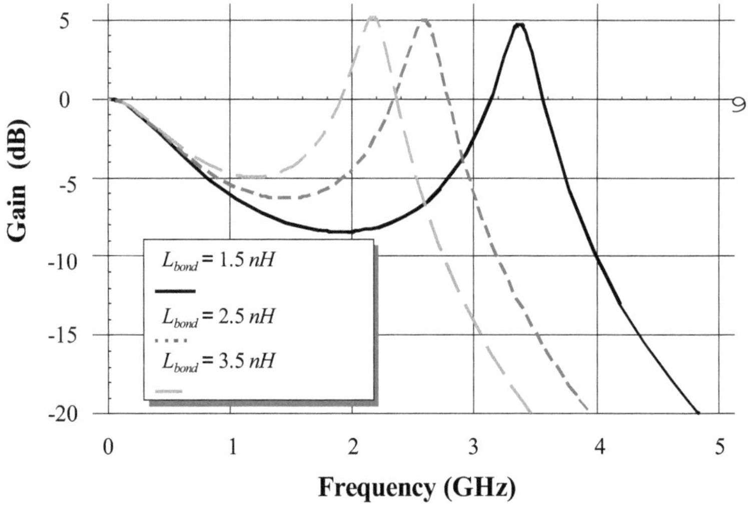


**Fig. D.2** Amplitude of the gain as a function of the frequency of the bond-wire parasitic circuit

The previous graph shows that there is a 5 dB peaking in the amplitude of the voltage gain, this frequency is related to the value of the bond-wire inductance. This means that PSD of the input signal will be amplified at the frequency value where the peaking occurs. This can result in ringing in the time domain. However the frequency where this peaking of the gain occurs is higher than the bandwidth of the expected signals, which is inferior to 1 GHz, therefore the value of the PSD of the signal should be small and it is unlikely that ringing will occur in the signal. It is important to notice that using a smaller bond wire increases the frequency value where the gain peaking occurs. If the silicon die is connected directly to the PCB board the bond-wires are smaller than 1.5 mm and the gain peaking frequency should be higher than 3 GHz. The parasitic circuit also attenuates the PSD of the signal within its band, but this attenuation is only 5 dB, therefore it is expected that this parasitic circuit should not influence significantly the gain of the overall circuit. The previous graph does not change significantly within the signal band, if the the values of $C_{pad}$ and $C_{pin}$ change 50% around the previous values.

# Appendix E
# Oscilloscope Characterization

In this appendix, the digital oscilloscope (Tektronix TDS3052) used to measure the output voltages of the circuit is characterized. This characterization allows to determine the non-ideal behavior of the oscilloscope and how this behavior affects the measurements made by the oscilloscope.

The digital oscilloscope is essentially a high speed sample and hold circuit followed by an ADC, therefore it introduces a quantization error and a thermal noise voltage into the sampled signal. In order to determine the values of these voltages, the output signal of the oscilloscope when no input signal is present must be analyzed. Ideally, when the input of the oscilloscope is shorted to the ground, the oscilloscope should measure an input voltage equal to 0 volt. After configuring the oscilloscope with a vertical resolution equal to 200 mV/div, a sampling frequency equal to 250 MHz and programming the oscilloscope to have a 20 MHz input low-pass filter in the signal path, 10000 differential voltage samples of the input signal are recorded. Each differential sample is the difference between the voltages acquired by each of the two channels in the oscilloscope. Each of these channels has its own ADC with a resolution of 9 bits. The oscilloscope calculated internally the difference between the two channel voltages. The differential output signal produced by the oscilloscope is shown in Fig. E.1.

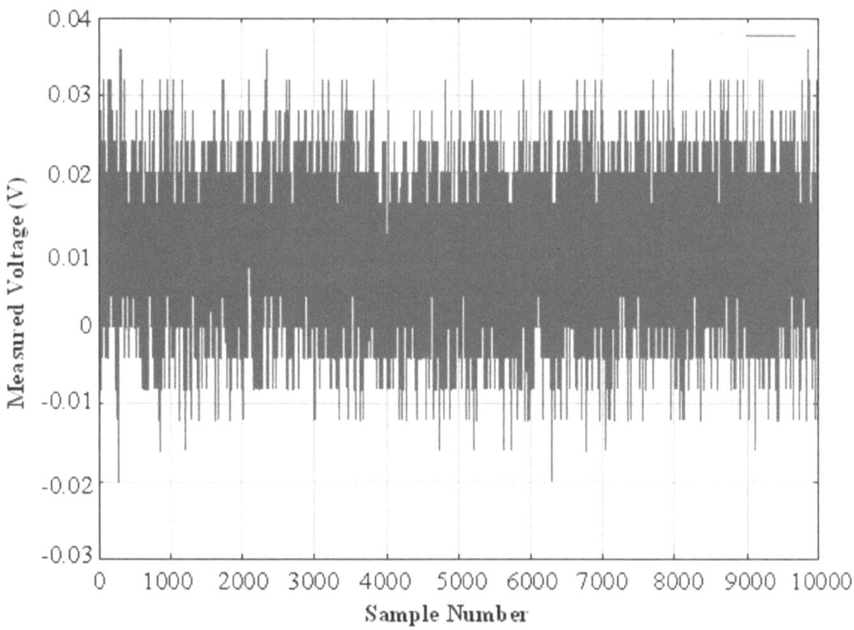


**Fig. E.1** Sampled differential voltage for a 0 V input signal

The differential output voltage produced by the oscilloscope is as expected a random signal; the noise power of this signal can be determined by calculating the standard deviation of the signal resulting in a noise voltage equal to 8 mVrms. The output signal also has an offset equal to 9.5 mV. The noise voltage in this signal is due to the thermal noise in the oscilloscope circuits and due to the quantization noise from the ADC. In order to determine the value of the quantization step of the oscilloscope ADC (*lsb* voltage) it is necessary to compute the histogram of the previous signal. This histogram must be calculated using a bin-width much smaller than the value of the *lsb* voltage in order to observe the quantization effect of the ADC. The histogram of the signal is shown in Fig. E.2. This histogram shows that the values of voltage samples produced by the ADC are quantized since they only occur at certain voltage bins. The voltage values are clustered around bins separated by 4 mV. This means that the *lsb* voltage of the ADC is equal to 4 mV. This corresponds to a quantization noise power equal to 1.15 mVrms. The thermal noise power of the oscilloscope can be calculated using $V_{NT} = \sqrt{V_N^2 + V_{NQ}^2}$ and is equal to 7.92 mVrms.

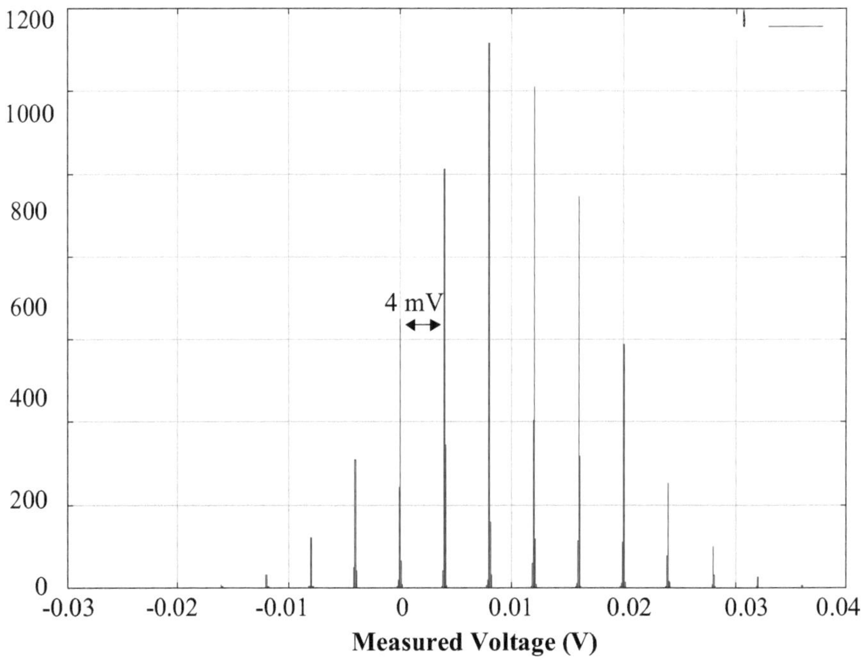


**Fig. E.2** Histogram of the sampled input voltage

When the ADC is used to measure a voltage signal in a test circuit, the resulting measured signal will have noise due to the circuit and due to the oscilloscope. If the noise power of the test circuit is much smaller than the noise power of the oscilloscope the noise power present in the measurement will be essentially due to the oscilloscope. In order to measure the noise power of the test circuit it is necessary to filter the noise power due to the oscilloscope. This can be achieved using averaging. By averaging a periodic signal its noise power can be reduced the number of averages. The oscilloscope allows to perform this operation internally. However, this procedure will also reduce the noise power from the test circuit itself and would prevent the measurement of the correct circuit noise power value (since it would also be reduced by the averaging operation). If the test circuit voltage is a sampled signal, there is an option to reduce only the oscilloscope noise power value and leave the test circuit noise power value unaffected. The test circuit voltage can be measured by the oscilloscope using a sampling frequency much higher than the signal sampling frequency. This way, each voltage sample produced by the circuit is sampled several times by the oscilloscope, corresponding to a set of samples that can be averaged to obtain a single voltage value corresponding to the test circuit sample. Since this set of samples correspond to the same test circuit sample, the averaging operation does not affect the noise power value of the circuit, which is a DC signal from the point of view of the averaging operation.

The filtered noise power value of the oscilloscope will depend on the number of samples averaged together. In order to determine these values the signal shown in Fig. E.1 is used to create signals where each sample corresponds to the average of $N$ samples of the original signal. The standard deviation of each of these signals is then calculated. Note that by grouping and averaging $N$ samples together, results in a signal witch has only $10000/N$ samples and therefore the standard deviation value might not be accurate when the value of $N$ is high. The noise power values of the oscilloscope obtained using different values for $N$ are shown in the table E.1.

**Table** E.1 Filtered oscilloscope noise voltage as a function of the number of samples added together

| N | Vrms |
|---|---|
| 1 | 0.00808302 |
| 10 | 0.00314001 |
| 20 | 0.00255002 |
| 50 | 0.00193409 |
| 100 | 0.00144772 |
| 200 | 0.00108061 |
| 1000 | 0.000707423 |

# Index

MIX
Papier aus verantwortungsvollen Quellen
Paper from responsible sources
FSC® C105338

If you have any concerns about our products,
you can contact us on
**ProductSafety@springernature.com**

In case Publisher is established outside the EU,
the EU authorized representative is:
**Springer Nature Customer Service Center GmbH**
**Europaplatz 3, 69115 Heidelberg, Germany**

Printed by Libri Plureos GmbH
in Hamburg, Germany